Presented to

**Spring Branch Memorial Library
In Memory of Vickie Blanks**

By

**John & Kathy Braddock & Family
Danny & Molly Lang & Family
Steve & Caroline Compton &
Family
Kevin & Susan Small & Family**

Harris County
    Public Library
      your pathway to knowledge

Intriguing Mathematical Problems

Oswald Jacoby

with William H. Benson

Dover Publications, Inc.
Mineola, New York

Bibliographical Note

This Dover edition, first published in 1996, is an unabridged republication of the work originally published by McGraw-Hill Book Company, Inc., New York, in 1962 under the title *Mathematics for Pleasure.*

Library of Congress Cataloging-in-Publication Data

Jacoby, Oswald, 1902–
      [Mathematics for pleasure]
      Intriguing mathematical problems / Oswald Jacoby with William H. Benson.
              p.   cm.
      Originally published: Mathematics for pleasure. 1st ed. New York : McGraw-Hill, 1962.
      ISBN 0-486-29261-4 (pbk.)
      1. Mathematical recreations.   I. Benson, William H.   II. Title.
QA95.J3 1996
793.7'4—dc20   96-19505
                                                                                    CIP

Manufactured in the United States of America
Dover Publications, Inc., 31 East 2nd Street, Mineola, N.Y. 11501

foreword

By the time I was twenty-one I had become a fair mathematician, an actuary, and a good cardplayer, and I have had some success in these fields in the years since. One factor that contributed to these accomplishments was my father's practice, starting when I was about six, of posing for me such problems as "If a ball and a bat cost a dollar and a half between them and the bat costs a dollar more than the ball, how much does each cost?"

As I grew older I became exposed to more complicated problems, some of which required mathematical knowledge and all of which demanded the use of reasoning. In the process I became an avid puzzle fan and an inveterate reader of books dealing with the lighter side of mathematics and logic.

Naturally enough, I began to want to produce a book of this sort myself—but one that would be different, challenging, and interesting. I wanted to offer a book that would avoid problems too easy, too tedious, or too involved in any great amount of formal mathematics.

Mathematics for Pleasure* is the result. In its preparation I had the help of the perfect co-author, William H. Benson, Associate

* Original title of the present volume.

Professor of Mathematics at Dickinson College—an old friend I had first known when he was Captain Benson, USN, and I was Commander Jacoby, USNR. In addition to his faculty for turning out highly accurate work of all sorts, Professor Benson has long made a hobby of mathematical recreations and ranks with the world's top experts on magic squares.

The book is made up of problems, with solutions grouped at the end of each section. These problems are selected for their interest and challenge and are much more than a casual collection of problems. Each one was carefully screened for its ability to stimulate imagination and interest and to allow you to put your reasoning powers to work. In very few is there a requirement for advanced mathematical knowledge or prowess, but in all of them you will have to keep your wits—your reasoning power—going at full steam. While the problems have been grouped according to type, they have not been arranged in any particular order of difficulty. So far as actual mathematics is concerned, the average person who hasn't used his mathematics since school will be surprised how much of what he learned in the past he still retains.

Some of the problems you will meet in this book are the work of Hubert Phillips (known also as Caliban) of London, who is unquestionably the world's leading problem composer. He has written more than nine books on mathematical problems and has conducted a well-known and popular British newspaper feature on problems of this general nature. He kindly made available to us his complete material.

In addition to Hubert Phillips' cooperation, we wish to acknowledge with deep thanks the assistance of Professor John H. Light of Dickinson College; and of Dr. Edward Cogan, Professor of Mathematics at Sarah Lawrence College, and Captain Wesley A. Wright, USN (Ret.), who carefully checked all problems, solutions, and other material.

Oswald Jacoby

contents

1. fun with numbers   1

1. the enterprising snail · 2. bargain day · 3. the bright graduate · 4. the erudite bookworm · 5. mother's little kitchen helper · 6. seven loaves of bread · 7. average speed · 8. the ladder · 9. remainder of one · 10. how much dirt? · 11. ten digits · 12. rare flowers · 13. chickens · 14. send more money! · 15. the empty flask · 16. mrs. crabbe and the bacon · 17. window-washing · 18. mr. spendmore · 19. time? · 20. tires · 21. trains · 22. prime settlement · 23. mixture

solutions: page 12

2. fun with letters   22

1. the professor's fish · 2. floating down the river · 3. one hundred dollars needed · 4. what is the product? · 5. joe nitwit · 6. the fly and the bicy-

cles · 7. birds in flight · 8. phil anthrope · 9. the wire fence · 10. my favorite orange · 11. candles · 12. three drain pipes take—? · 13. helen's age is? · 14. jack and jill · 15. mrs. crabbe and the prunes · 16. trust fund · 17. what is the first time? · 18. round trip from abletown · 19. mr. wright and the insurance salesman · 20. philip gibson's children · 21. ducks · 22. ten-o'clock scholar · 23. packing crate · 24. trains passing · 25. wheat · 26. the professor's window · 27. as the crow flies · 28. courier · 29. outboard *vs.* speedboat   **solutions: page 34**

3. the odds: explorations in probability   53

1. cue's chance · 2. transfers · 3. the odds are two to one · 4. how many left? · 5. tennis at lower dropshot · 6. face cards · 7. the king of hearts · 8. birthdays · 9. the walk to 15th and m · 10. the three prisoners · 11. the dictator · 12. more tennis at lower dropshot · 13. cubes · 14. king's town · 15. professor of ancient history · 16. two spades · 17. smithley's problem · 18. the odds against smith · 19. professor knowsodd's reply · 20. opinionated is not satisfied · 21. a pair of dice · 22. girls should live in brooklyn · 23. the three chess players · 24. the five balls · 25. how many balls?   **solutions: page 66**

4. where inference and reasoning reign   84

part 1. inference   87

1. ho island · 2. the island of ha · 3. bridge · 4. the diners · 5. "fur and feather" show · 6. the dine-out club · 7. who was executed? · 8. literary dinner party · 9. bridge at the greens' · 10. mr. smith is? · 11. the squabble · 12. the new member cut in · 13. bella's gifts · 14. smithley back at ho island · 15. bridge on ha island · 16. confusing? · 17. flowers · 18. king arthur · 19. outdoors · 20. horsemen · 21. muddle at mixwell · 22. five sons · 23. the island of hi · 24. david? · 25. the island of hu · 26. artichoke, bergamot, and celery · 27. the bank manager is? · 28. white and green · 29. strange!   **solutions: page 100**

part 2. reasoning   121

30. the light coin · 31. how often does christmas fall on friday? · 32. the lovesick cockroaches · 33. chain smoker · 34. commuter service · 35. beefeater's convention · 36. the walking fly · 37. chessboard · 38. u.s. coins · 39. white hats · 40. bigdome's park · 41. jim and judy · 42. jones was early · 43. north · 44. the prince · 45. creaker *vs.* roadhog · 46. chains · 47. triangles · 48. as the tide rises · 49. what time is it? · 50. the fresh-air fiend · 51. mr. jones · 52. billiard plays? · 53. much dithering · 54. the back-alley gang · 55. the seven housemasters · 56. his instructor's age · **solutions: page 135**

5. the answers are whole numbers

1. the farmers' market · 2. jody and sandra · 3. cashier's error · 4. the crossed ladders · 5. egg money · 6. wendy, sally, and debby · 7. monkey business · 8. scholarships · 9. birthday cakes · 10. which is nearer? · 11. muggins · 12. mary · 13. knights *vs.* bishops · 14. the price went down · 15. four brothers and their gin game · 16. the ancient order of the greens · 17. mrs. overtwenty · 18. how many men? · 19. seventeen pencils · 20. jones' children · 21. mrs. quigley · 22. the *rhind papyrus* · 23. poppy · 24. the graustark cabinet · 25. the two cousins · 26. professor algebra's garden · 27. coconuts · 28. a box of chocolates · "seventeeners"   **solutions: page 167**

Intriguing
Mathematical
Problems

fun with numbers 1

When man first began to count—probably well before he began to write—it was only natural that he use the facilities he had. There were five fingers on each hand, so at first he counted to five and not long after that to ten. If he wanted to move into the realm of higher numbers he could use his ten toes and get as far as twenty. After that he had to start all over again—which is why almost every language has a single-word name for each number up to twenty but starts with double and triple words from then on.

The use of the toes for counting never became very popular, even though man was counting nicely before he stopped going barefoot. It was easier to think of higher numbers as so many tens and units until the important number of ten tens was reached. A name was needed for this, and still another for ten tens of tens, so from very early times most languages have had words or symbols for *hundred* and *thousand*.

Written numeration (in its broadest sense) goes back to pre-historic man. The cuneiform baked-clay tablets of the Sumerians and the Babylonians, as well as Egyptian records on papyrus,

1

prove that as long as five thousand years ago these peoples had a usable system of numbers. The Egyptian system is considered a decimal system: it had names and symbols for 10, 100, 1000, 10,000, 100,000, and 1,000,000; it did not, however, employ the principle of position or use a zero.

The Sumerians and Babylonians used a sexagesimal or 60-base system that did include the principle of position, and late in the game (about 200 B.C.) introduced a symbol for zero.

By the time they had reached the height of their culture, the Mayas of Central America had a calendar and a highly developed 20-base system of numbers that included zero and the principle of position.

The Greek system used the entire Greek alphabet as a series of symbols representing numbers in arithmetical operations. The Roman system, which also used specific letters and letter combinations to represent numbers, was an improvement but remained cumbersome and inefficient. Even with the abacus, which made addition and subtraction feasible, multiplication and division as we know these processes were not possible. Before arithmetic could gain any real momentum a better system was necessary.

During the early centuries of the Christian era the Hindus of India furnished the solution by formulating the principle of position and by the use of ten symbols to represent zero and the first nine digits. With these two developments, larger numbers could be expressed simply in terms of the position of each symbol. Thus 24 represents two tens and four units; 2304 represents two tens of tens of tens plus three tens of tens plus four units, or

$$2(10)^3 + 3(10)^2 + 0(10) + 4.$$

Despite the immediately obvious advantages of this system, not until around the beginning of the thirteenth century did it reach Western Europe (by way of Arabia), and its adoption there was not more or less complete until the beginning of the sixteenth century.

The system, good as it was, still left a problem with fractions— a problem that was solved by the introduction during the late sixteenth and early seventeenth centuries of the decimal system. After the decimal system began to be used, it became simple to represent two tens plus eight units plus one tenth plus six hundredths, for example, merely by writing 28.16. Just as each place to the left of the decimal point represents a multiplication by ten, so each place to the right represents a division by ten.

Both the Romans and the Egyptians had a word or symbol for *ten, hundred,* and *thousand;* we have words for much larger numbers. *One million* is one thousand thousands and is written 1,000,000; *one billion* is one thousand millions and is written 1,000,000,000; and *one trillion* is one thousand billions and is written 1,000,000,000,000. While Arabic numerals and the decimal point take care of reasonably large and reasonably small numbers, this does not mean that we can understand or visualize what a large number stands for. For instance, 376,987,543,843,217 is a fifteen-digit number. It can be referred to as three hundred seventy-six trillion, nine hundred eighty-seven billion, five hundred forty-three million, eight hundred forty-three thousand, two hundred seventeen. Beyond that its meaning is questionable. We also have words for even larger numbers, but unless we can express these larger numbers in terms of smaller numbers, they mean nothing to us.

And today we have increasing need for large numbers. Consider these: There are 31,536,000 seconds in one year; if a man could count at the rate of ten per second, it would take him more than three years to count to a billion. Astronomers have to deal with vast numbers, since the known universe covers a considerable amount of space. In order to simplify matters they use a unit known as the *light year,* the distance light travels in one year. Since light travels at the rate of a trifle over 186,000 miles per second, the number of miles in a light year is a thirteen-digit number slightly smaller than six trillion. Our national debt is a twelve-digit number, but our national wealth is conservatively estimated to be a thirteen-digit number. And 635,013,559,600 is a fairly well-known twelve-digit number; it represents the number of possible bridge hands a player can hold.

Most of the classic problems that deal with large numbers are concerned with the fact that when you multiply enough small numbers together you come up with a very large number.

Tradition has it that the inventor of the game of chess was invited by his emperor to name his own reward. "All I ask," the inventor is supposed to have said, "is to be awarded one grain of wheat for the first square on the chessboard, two grains for the second, four for the third, eight for the fourth, and so on. His Majesty laughed heartily and agreed at once to the apparently modest request. He changed his mind around the twenty-first square and had the man hanged. The emperor had to come up with 1,048,576 grains of wheat for that square and, although it was within his limits, he realized he would before too long exhaust his wheat resources. Just in case you don't know how many grains of wheat would eventually have been required, the answer is $2^{64} - 1$, or 18,446,744,073,709,551,615.

But for large numbers or small, when you solve the problems that follow in this section of the book, only the principles of arithmetic that are known to all of us will be involved—with one possible exception: the concept of *prime numbers*. If you consider the basic arithmetical operations of addition, subtraction, multiplication, and division when applied to whole numbers (or *integers*), you will notice immediately that the sum, difference, or product of any two integers is also an integer. However, if you divide one integer by another, the result is not necessarily an integer; in fact, it is not likely to be an integer unless you have carefully chosen both divisor and dividend with this end in view.

Thus if you try dividing by 7 you will find that 7, 14, 21, 28, and so on are exactly divisible by 7, while all the intermediate integers are not. 8 is exactly divisible by 1, 2, 4, and 8; 9 by 1, 3, and 9; 12 by 1, 2, 3, 4, 6, and 12. Such divisors are known as *factors* of the numbers they divide exactly.

The integers 2, 3, 5, 7, 11, and all other numbers which contain no factors other than themselves and unity, are known as *prime numbers*. You will find of particular interest in several of the problems in this section the Fundamental Theorem of Arithmetic, which states that "Except for order, any integer can be resolved into a product of primes in only one way."

If some of the problems in this section seem too simple to you, don't worry—and read carefully! The problems throughout the book are calculated to challenge you. Much of that challenge lies in the fact that logic and clear-cut reasoning will be of much more value to you than extensive mathematical know-how, even in solving the ones that sound simplest.

1. the enterprising snail

A group of snails had been living happily for many years at the bottom of a twenty-foot well when it suddenly ran dry. Most of the snails lay down to die, but one industrious fellow decided to go elsewhere. The next dawn he began to climb to the top of the well at the prodigious rate of three inches per hour for each of the twelve daylight hours. Each night, however, he slipped back at the rate of two inches per hour. How long did it take this enterprising snail to escape from the well?

2. bargain day

On the final day of his close-out sale, a merchant hastily disposed of two lamps at the bargain price of twelve dollars apiece. He estimated that he must have made some profit on the combined transactions since he made a 25 per cent profit on one and only took a 20 per cent loss on the other.

Was he correct in his quick estimate?

3. the bright graduate

Young Smitherly, who was about to graduate at the head of his class at Upcoming University, was in the pleasant position of having his choice of two very attractive offers—both at $5000 a year. Unable to make up his mind between them immediately, he wrote the two companies and asked what his chances were for raises over the next five years.

Company *A* replied to the effect that it would guarantee a raise of $300 every six months for the next five years. Company *B* said it would guarantee a raise of $1200 every twelve months for the next five years.

To the astonishment of his father, Smitherly accepted Company *A*'s offer. Why?

4. the erudite bookworm

The bookworm is a staple character in puzzle literature. Ours, finding himself in a library in which the sets of books were placed in the usual sequence, decided to sample some Shakespeare from a two-volume edition on a bottom shelf. Beginning with the Foreword of Volume 1 and boring through in a straight line to the last page of Volume 2, the bookworm made his way at the rate of one inch every four days. If each cover is an eighth of an inch thick, and if each volume measures three inches in thickness, how long will it take the erudite bookworm to digest his way through Shakespeare?

5. mother's little kitchen helper

If it takes 3¾ minutes to boil one egg, how long does it take to boil six eggs?

6. seven loaves of bread

Three travelers meet. They sit down to eat. One traveler produces three loaves of bread, the second traveler produces four loaves. After the seven loaves are divided equally and eaten, the third traveler produces seven dimes and says, "Please divide these dimes equitably between the two of you." How much money should each of the first two travelers receive?

7. average speed

Mr. Sandys took a great deal of satisfaction from the fact that his summer cottage was comparatively close to his office in the city. One Friday afternoon he got an early start and was able to make the trip to the cottage at an average speed of sixty miles an hour. But he overslept on Monday morning and by the time he got on his way to the city the traffic was so heavy that he averaged only thirty miles an hour. By the time he reached his office, Mr. Sandys was weary and much less satisfied with the convenience of his week-end retreat.

What was his average speed for the round trip?

8. the ladder

At a factory fire a fireman stood on the middle rung of a ladder, directing water into the burning building. As the smoke diminished, he stepped up three rungs and continued his work from that position. A sudden flare-up of flames forced him to descend five rungs. A few minutes later he climbed up seven rungs and worked there until the fire was out. Then he climbed the seven remaining rungs and entered the factory.

How many rungs were in the ladder?

9. remainder of one

What is the smallest number divisible by 13 which, divided by any of the numbers from 2 to 12 inclusive, leaves a remainder of 1?

10. how much dirt?

How many cubic inches of dirt are there in a hole that is one foot deep, two feet wide, and six feet long?

11. ten digits

1. Arrange the digits 1, 2, 3, 4, 5, 6, 7, 8, and 9 in succession using each one only once and, with the aid of plus and minus signs as desired, produce a sum of 100.

2. Add zero to the nine digits above and rearrange them in fraction form to produce a sum of 1.

12. rare flowers

It was a day of special joy for Mrs. Greenhouse. She was exhibiting her collection of rare flowering plants to members of her garden club.

"Now these are my real treasures," she said in her best lecture voice, pointing to three luxuriously flowering specimens. "I got them from Guatemala only a year ago, and I have been giving them my closest attention. They are all quite delicate indeed, since the flowers last only a single day. The plant with crimson flowers blossoms every fourth day, but I have to wait until the seventh day for the plant with lavender flowers. But that one is the most precious of all," she said, directing attention to her third plant. "It takes a full thirteen days for it to blossom."

Since her three tropical flowering plants were such an enormous attraction, Mrs. Greenhouse expansively invited the club members to return to see them again. "Let's make it next year on this very same day," she announced, "since they will all be in blossom then."

Was Mrs. Greenhouse correct in her horticultural reckoning?

13. chickens

If a hundred chickens eat a hundred bushels of grain in a hundred days, how many bushels will ten chickens eat in ten days?

And if, on the average, one and a half of these chickens lay an egg and a half in a day and a half, how many days will it take a chicken to lay one and a half dozen eggs.

14. send more money!

A father, much irritated by his undergraduate son's frequently repeated demands for more money, rebelled the day he received one in the same mail with the Registrar's report of his son's poor grades.

The next letter from the father was brief and to the point:

Dear Son,

If you won't do any work for your professors maybe you will for me! At any rate there will be no money forthcoming until you send me the solution to the following:

$$\begin{array}{r} S\,E\;N\;D \\ +\,M\,O\;R\;E \\ \hline M\;O\;N\;E\;Y \end{array}$$

As a hint, each letter represents a different digit.

Love,
Dad

Can you help the son solve it?

15. the empty flask

An apothecary found six flasks capable of holding sixteen, eighteen, twenty-two, twenty-three, twenty-four, and thirty-four fluid ounces respectively. He filled some with distilled water and then filled all but one of the rest with alcohol, noting that he had used twice as much alcohol as water.

Which flask was left over? And which flasks were used for distilled water, which for alcohol?

16. mrs. crabbe and the bacon

At Harmony House it is Mrs. Crabbe's custom to set up a platter of breakfast bacon so that her lodgers may help themselves as they come down in the morning. On this particular morning, had all the lodgers shared alike, there would have been a whole number of rashers each, but the Oldest Boarder, who came down last, found only half a rasher left for him. Jones always takes one rasher; Robinson takes his fair share of what remains; Brown is greedy, takes his fair share and then half a rasher extra; Wag likes three rashers, but, being superstitious, always leaves at least one for those who follow him.

How many rashers did Wag take?

17. window-washing

When Mr. Watkins built his home, he made one wall of the living room entirely out of thick plate glass. He soon realized how expensive—as well as inconvenient—it was to live in a "glass fishbowl," for the glass had to be washed on the average of once a week. It was too big a job for him, since the wall measured nineteen by twenty feet, so he contracted with a window-cleaning service to do the full job once a week at the cost of one dollar per hundred square feet of surface washed.

What was Watkins' weekly bill?

18. mr. spendmore

When Mr. Spendmore counted the money in his pocket the "morning after," he found a single crumpled dollar bill. Ruefully he recalled the hectic evening he had spent on the town, weaving a trail from one night spot to the next.

He tried to remember exactly how much money he had with him when he started his gala evening. But all he could remember was that he had spent half his money at the Top Hat, his first stop, and that as he left he had tipped the hat-check girl a dollar bill. At the Golden Eagle, the second night club, he had spent half his remaining money and again had tipped the hat-check girl a dollar bill. He repeated the same methodical performance at the Glass Slipper and again at the Pirate Ship before he finally staggered home.

How much did the extravagant Mr. Spendmore have when he started out?

19. time?

Simpkins and Green made arrangements to meet at the railroad station to catch the eight-o'clock train to Philadelphia. Simpkins thinks that his watch is twenty-five minutes fast although it is in fact ten minutes slow. Green thinks his watch is ten minutes slow, while it actually has gained five minutes.

What will happen if both men, relying upon their watches, try to arrive at the station five minutes before train time?

20. tires

The MacDonalds were a young, adventuresome couple. With a great deal of foresight and calculation, they made elaborate plans for a summer tour in their battered station wagon through the less-traveled scenic parts of Mexico and Central America. Mapping their itinerary in advance, they calculated that their journey would cover a total of twenty-seven thousand miles.

On the advice of a friend, the MacDonalds decided to buy special tires, each one guaranteed to last twelve thousand miles provided the car was not overloaded. Jettisoning the heavy equipment so they would not load the station wagon with more weight than was allowed by the guarantee, they now calculated how many tires they would need for this long-distance trip.

What is the least number of tires that would carry them through their journey, and how could they make the best use of them?

21. trains

A train leaves New York for Washington every hour on the hour. A train leaves Washington for New York every hour on the half-hour. The trip takes five hours each way.

As you ride from New York to Washington, how many of the trains bound from Washington to New York would you pass?

22. prime settlement

The Saturday-night card game was over and the books balanced. The big loser commented, as he placed his check on the table, "There goes a prime number of dollars." The only other loser paid his debt with four bills of US currency of different denominations, commenting as he placed them on the table, "And here's another, but I'm glad to say smaller, prime number of dollars."

The biggest winner picked up the check and placed a hundred-dollar bill—the difference between the amount he had won and the big loser's check—on the table and said, "Well, I've got my prime number of dollars and there's still a prime number of dollars for the rest of you to divide." Surprisingly enough, as each of the five remaining men took his share in order—the second biggest winner first, the third biggest winner next, and so on—each took a single bill as his share and commented as he did so, "Well, I wouldn't think of disturbing the prime quality of the amount remaining."

Assuming all the men used the word *prime* in the mathematical sense (a number having no integral divisors other than unity and itself), what is the smallest amount that could have been won and lost and still meet the stated conditions?

23. mixture

One glass is half full of wine and another glass twice its size is one-quarter full of wine. Both glasses are filled with water and the contents mixed in a third container. What part of the mixture is wine?

solutions

1. the enterprising snail

17½ days. The snail climbed at the net rate of one foot per day (3 − 2 = 1). At dawn of the start of the eighteenth day he was, therefore, 17 feet from the bottom or 3 feet from the top. Since he climbed 3 feet during the twelve daylight hours of the eighteenth day he would reach the top at sunset of the eighteenth day, or 17½ days after he started his climb.

2. bargain day

No.

$12.00 is 125% of $9.60   — $2.40 profit
$12.00 is 80% of $15.00   — $3.00 loss
                                ―――――――――
                              $ .60 net loss

3. the bright graduate

During each year Smitherly would receive the following from the
two companies:

Year	Company A	Company B
First	$2500 + $2800 = $5300	$5000
Second	$3100 + $3400 = $6500	$6200
Third	$3700 + $4000 = $7700	$7400
Fourth	$4300 + $4600 = $8900	$8600
Fifth	$4900 + $5200 = $10,100	$9800

In other words, Company A's offer is $300 a year better than
Company B's.

This looks like a paradox but it isn't. A raise in semiannual
pay of $300 is the same as a raise of $600 in annual pay. Since
Company A was offering two such raises, or $1200 a year, it was
equivalent to Company B's offer except that, as a result of the
semiannual feature, each year he would get $300 of this $1200
raise six months earlier.

4. the erudite bookworm

Since this library is arranged in usual order, Volume 1 will be
to the left of Volume 2. The Foreword of Volume 1 and the last
page of Volume 2 will be separated only by two covers—a distance of ¼ inch. At his specified rate of travel, the bookworm will
munch his way through this distance in one day.

5. mother's little kitchen helper

3¾ minutes, provided you boil them all at the same time.

6. seven loaves of bread

The key is the word *equitably*. At first glance it may appear that the first traveler should receive three dimes and the second should get the remaining four, but this would not be equitable. Each traveler has eaten ⅓ loaves of bread. The first traveler contributed $(3 - ⅓) = ⅔$ loaves to the third traveler and the second traveler contributed $(4 - ⅓) = ⅗$ loaves to the third traveler. For the distribution to be equitable, the first traveler should receive two dimes and the second traveler five dimes.

7. average speed

Forty miles an hour. This figure holds true no matter what the distance may be. For example, assume that the trip from city to summer cottage covered 120 miles. On the trip out to the country Mr. Sandys averaged 60 miles an hour, thus taking two hours to cover the distance. Since he made the trip back to the city at an average speed of 30 miles an hour, he took four hours over the same distance. In all, the round trip of 240 miles took him six hours, thus giving him the 40-mph average.

8. the ladder

If the middle rung is marked 0, the fireman went up three rungs to number 3, down five rungs to number 2 below 0, then up seven rungs to number 5 above. Finally he mounted seven more rungs to the top. So the top rung must have been number 12 above the middle. Add to this the twelve rungs below the middle and the middle rung itself. The ladder had twenty-five rungs.

9. remainder of one

The least common multiple of the numbers 2 to 12 inclusive is 27,720. The required number is, therefore, of the form $27,720n + 1$. Dividing by 13 gives

$$\frac{27,720n + 1}{13} = 2132n + \frac{4n + 1}{13}$$

If the number is to be divisible by 13, $4n + 1$ must be a multiple of 13. Obviously $n = 3$ is the smallest integral value of n that will satisfy this condition.

The required number $= 3 \times 27{,}720 + 1 = 83{,}161$.

10. how much dirt?

None. The dirt has been taken out.

If you want to know the number of cubic inches of dirt taken out the answer is $1728 \times 6 \times 2 \times 1 = 20{,}736$ cubic inches.

11. ten digits

1. $$123 - 45 - 67 + 89 = 100$$

2. $$\frac{35}{70} + \frac{148}{296} = \tfrac{1}{2} + \tfrac{1}{2} = 1$$

The first answer appears to be unique, but there are numerous ways to rearrange the second answer.

12. rare flowers

If the club members come a year later on the very same day of the month, they would be one day late for the triple blooming. On the other hand, if they come next year on the very same day of the week (and as nearly as possible a year later), they would be right on time.

The least common multiple of 4, 7, and 13 is the smallest number divisible by all three numbers. Since this number is $4 \times 7 \times 13 = 364$, it can be seen that the three plants blossom simultaneously every 364 days—52 weeks, but one day less than a full year.

13. chickens

The tendency in the first problem is to conclude that if 100 chickens eat 100 bushels in 100 days, one chicken will eat one bushel in one day—and any farmer will certainly quarrel with the likelihood of this calculation. The correct reasoning leads to this conclusion: If any number of chickens eat 100 bushels in 100 days, then the *same number* of chickens will eat one bushel in one day. This reasoning yields the correct answer: 10 chickens will eat one bushel in 10 days.

The same process of reasoning leads to 27 days as the time required for one chicken to lay 1½ dozen eggs.

14. send more money!

M is obviously 1, since there is no way there can be a greater carryover than one when only two digits are added. Similarly, O must be zero and S either 8 or 9. A quick check shows that if $S = 8$, E must equal 9, and N will equal zero. But this is impossible because $O =$ zero. So $S = 9$, making the result so far:

$$
\begin{array}{r}
9\;E\;N\;D \\
+\quad 1\;0\;R\;E \\
\hline
1\;0\;N\;E\;Y
\end{array}
$$

Since E plus zero plus carryover equals N it follows that

$$N = E + 1$$

and that N plus R is greater than 9 (there must be a carryover), or

$$N + R = E + 10$$

if there is no carryover from the first column; if there is a carryover

$$N + R = E + 9.$$

Substituting for N its equivalent value $(E + 1)$ in these last two equations gives

$$E + 1 + R = E + 10 \quad\text{and}\quad E + 1 + R = E + 9$$

or $R = 9$ or 8

and, since $S = 9$, R must equal 8 and $E + D = 12$ or more (there must be a carryover and the digits 0 and 1 have already been assigned). Therefore, the only possible values for E, D, and

N are 5, 6, and 7. Remembering that $N = E + 1$ and $E + D = 12$ or more, it follows that

$$E = 5 \qquad N = 6 \qquad D = 7 \qquad \text{and} \qquad Y = 2$$

All the letters are now identified and the addition becomes

$$\begin{array}{r} 9\,5\,6\,7 \\ +\;\; 1\,0\,8\,5 \\ \hline 1\,0\,6\,5\,2 \end{array}$$

15. the empty flask

Since the apothecary had used twice as much alcohol as water, the total ounces of alcohol and water together must be divisible by 3. The total capacity of all 6 flasks is 137 ounces, a number which is 2 more than a multiple of 3. Because the apothecary left one flask empty, it is obvious that this flask must have had a capacity in ounces which is 2 more than a multiple of 3. The only flask which meets this requirement is the one with a capacity of 23 ounces. This flask must have been the one left empty. The remainder, which have a combined capacity of 114 ounces, were filled. One-third of this total, 38 ounces, was water. The 16- and 22-ounce flasks must therefore have been used for water and the 18-, 24-, and 34-ounce flasks used for alcohol.

16. mrs. crabbe and the bacon

1. There are five lodgers; the number of rashers originally on the dish must be 5, or 10, or 15, or ____

2. The number is 5. The greatest possible original number would be reached in the following way:

The Oldest Boarder takes $\frac{1}{2}$ rasher; Jones takes 1, total $\frac{3}{2}$; Wag takes 3, total $\frac{9}{2}$; and then either

a. Brown takes $\frac{13}{6}$, total $\frac{20}{3}$; Robinson takes $\frac{5}{3}$, total $\frac{25}{3}$, or

b. Robinson takes $\frac{3}{2}$, total 6; Brown takes $\frac{17}{8}$, total $\frac{65}{8}$.

But both of these are less than 10; and the original number of rashers had to be five.

3. Since Wag always leaves at least one rasher, he cannot immediately precede the Oldest Boarder. There are three possibilities:

a. Jones precedes the Oldest Boarder and takes 1 rasher; total ⅜.

b. Brown precedes the Oldest Boarder and takes ⅜ rashers; total 2.

In both of these cases Wag must take 3 rashers, leaving only ½ or 0 for the other two; this is impossible.

c. Thus Robinson precedes the Oldest Boarder and takes ½ rasher; total 1.

Wag must precede Robinson; if not he must take three rashers, Jones must take one rasher, and there will be none left for Brown. It follows that

d. The order in which the lodgers come to breakfast must be
(1) Jones, Brown, Wag, Robinson, the Oldest Boarder, or
(2) Brown, Jones, Wag, Robinson, the Oldest Boarder.

e. So far as Wag is concerned, these two orders are the same.

If Brown comes first his fair share is 1, so he will take 1½ rashers. Jones taking, of course, 1 rasher, makes their total 2½ rashers and Wag finds 2½ rashers on the plate when he enters the breakfast room. If Jones comes first and takes 1 rasher, Brown will still take 1½ rashers because his fair share is not affected by Jones having taken a fair share before him. The result is that Wag again finds 2½ rashers when he enters. Hence we have: Brown and Jones 2½; Wag 1½; Robinson ½; the Oldest Boarder ½. Wag took 1½ rashers of bacon.

17. window-washing

The "full job" would have to include both the inside and outside —2 times 19 times 20 equals 760 square feet of surface to be washed. The bill was thus $7.60 per week.

18. mr. spendmore

Puzzles of this kind are most easily solved by working backward. If Mr. Spendmore had $1 at the conclusion of the evening, then he must have had $2 before he tipped the hat-check girl at the Pirate Ship and $4 before he paid the bill there. Moving backward, you can easily deduce that he had $4 when he left the Glass Slipper, $5 before he tipped the hat-check girl there, and $10 before he paid his bill. By repeating these calculations twice more, you find that Mr. Spendmore had $46 with him at the beginning of the evening.

19. time?

Simpkins thinks he will be in plenty of time when his watch shows 8:20, but it is already 8:30, much too late to catch the train. But Green will reach the station when his watch shows 7:45, and since his watch is five minutes fast, he will really be there at 7:40, twenty minutes ahead of train time.

20. tires

Nine tires are needed. A station wagon has four wheels, of course, so that for a 27,000-mile journey the car uses up 108,000 tire-miles. Since each tire is guaranteed for 12,000 miles, nine tires will mathematically satisfy the conditions.

But the calculating MacDonalds also realized the problem involved in making the best use of these nine tires. The first four tires will be worn out and discarded at the end of 12,000 miles. One of the five remaining tires would have to be changed each 3000 miles, according to this schedule:

First   3000 miles: Tires 1, 2, 3, 4
Second 3000 miles: Tires 2, 3, 4, 5
Third   3000 miles: Tires 3, 4, 5, 1
Fourth 3000 miles: Tires 4, 5, 1, 2
Fifth   3000 miles: Tires 5, 1, 2, 3

In this way each of the nine tires would be used no more than 12,000 miles.

21. trains

You would pass the first train returning to New York fifteen minutes after you start (you are fifteen minutes out of New York and the other train is due in New York fifteen minutes later) and one each half-hour thereafter.

You will pass ten trains bound for New York.

22. prime settlement

Under the conditions of the problem we know that the second loser paid a prime number of dollars with four bills of different denomination, say a, b, c, and d. As a first trial, let us assume that he lost less than $100 and that the smallest of the four bills was a, the next larger b, the next larger c, and the largest d. If we can find sets of values for a, b, c, and d, so that

a is a prime number
$a + b$ is a prime number
$a + b + c$ is a prime number
$a + b + c + d$ is a prime number
$a + b + c + d + 100$ is a prime number

we will, obviously, have a basis for a solution which is less than any solution which would be obtained under an assumption that the second loser lost more than $100. The sum of the three smallest-denomination bills is not a prime number ($1 + 2 + 5 = 8$). It follows, therefore, that c must be 10 or some multiple of 10 (there are no US bills larger than $5 which are not some multiple of $10). This leaves the following as the only possibilities for a and b (remembering that $a + b$ must be a prime number):

a	b	$a + b$
1	2	3
1	10	11
2	5	7

Trying out these possibilities gives us (numbers in parentheses are not prime and indicate that the particular values selected for a, b, c, and d, will not work):

a	b	c	d	a	$a+b$	$a+b+c$	$a+b+c$ $+d$	$a+b+c$ $+d+100$	all primes
1	2	10	20	1	3	13	(33)	(133)	no
1	2	10	50	1	3	13	(63)	163	no
1	2	20	50	1	3	23	73	173	yes
1	10	20	50	1	11	31	(81)	181	no
2	5	10	20	2	7	17	37	137	yes
2	5	10	50	2	7	17	67	167	yes
2	5	20	50	2	7	(27)	(77)	(177)	no

There are thus three possible solutions smaller than $100, and our trial assumption was justified. Since the smallest of these three solutions is $a = 2, $b = 5, $c = 10, and $d = 20, the conditions of the problem show us that the second loser lost $37 (the sum of $2 + 5 + 10 + 20 = 37$) and the smallest winner won $2, the next larger winner $5, the next $10, the next $20, and the next $100. It remains only to determine the amount won by the largest winner and the amount lost by the largest loser.

We know that these two amounts differ by $100 and that they are both prime numbers. Hence we must find the smallest prime number greater than 100 (the largest winner must have won more than $100) which remains prime when increased by 100. This number is 127. It follows that the biggest winner won $127 and the biggest loser lost $227 and that the amount which changed hands was $264 ($227 + 37 = 264$ lost, and $2 + 5 + 10 + 20 + 100 + 127 = 264$ won).

23. mixture

The two glasses hold a total of three units. The one-unit glass is one-half full of wine; the two-unit glass is one-quarter full of wine. Thus each glass holds one-half unit of wine, making the total amount of wine one unit. When the glasses are filled with water there will be two units of water and the mixture will be one-third wine.

fun with letters

If you like to mystify your friends with your mathematical prowess, bring a few of them together and ask each one to think of a number, any number. Addressing yourself to the entire group, you give a few simple arithmetic instructions. Soon you hand each person a piece of paper on which you have written the number he has just computed. You are, without a doubt, a mathematical whiz! For example:

Think of a number! Add 2! Multiply by 3! Add your original number! Multiply by 2! Add 2! Subtract your original number! Divide by 7! Add 3! Subtract your original number! Add 5! The result is 10 in all cases. It looks amazing, particularly if you keep varying your instructions from one example to another. All you are doing, however, is carrying the unknown in your head by letting x represent it. Then the successive operations above would give you x; $x + 2$; $3x + 6$; $4x + 6$; $8x + 12$; $8x + 14$; $7x + 14$; $x + 2$; $x + 5$; $x + 5 - x = 5$; $5 + 5 = 10$!

Note that the last operation with the original number removed it completely and everyone is left with 5. From there on you could have added as many steps as you wished, but all would have the same answer.

22

You used algebra while the rest of the group used simple arithmetic. The basic difference between algebra and arithmetic is that in algebra you apply the fundamental operations of arithmetic not only to numbers alone but also to numbers and to letters which represent numbers. Of course, just as seven horses and five cows are twelve animals but not twelve horses or twelve cows, so $7x + 5y + 8$ represents seven of one thing plus five of another thing plus eight.

Similarly, if you multiply $(a + b)$ by $(c + d)$, your answer will be $ac + ad + bc + bd$, while if you multiply $(a + 3)$ by $(a + 4)$ your answer will be $a^2 + 3a + 4a + 12 = a^2 + 7a + 12$. Likewise, $4^2 = 4$ times $4 = 16$, but a^2 merely equals a times a.

Suppose you write $4 + 3 = 7$. This is an equation and it is obviously correct. If you write $a + b = c$ you will also have an equation and it will be correct, provided you assign appropriate values to a, b, and c. Suppose you start by assigning values to a and b only. Whatever values you assign to them will determine the specific value you must assign to c for the equation to be correct. Thus when $a = 4$ and $b = 3$, c must equal 7. You can perform any arithmetical operation (other than dividing by a quantity equal to zero) to both sides of an equation and it will remain an equation. Thus if $a + b = c$, then $2a + 2b = 2c$; $a = c - b$; $a + b + 4 = c + 4$; and so forth.

Consider the two equations

$$2a + 2b = 2c \quad \text{and} \quad 2a + b = 2c.$$

If you subtract the second equation from the first you have

$$b = 0$$

and substituting this value in either of the original equations gives

$$2a + 0 = 2c \quad \text{or} \quad a = c.$$

You thus see that a necessary condition for these two equations to be correct is for a to equal c and for b to equal zero. Under normal conditions you cannot solve a group of equations when there are more unknown quantities than equations. An interesting exception to this general rule occurs in the case where the unknowns must be whole numbers, or integers. This is discussed in Section 5, *The Answers Are Whole Numbers.*

Be reassured that a high level of algebra is not required to solve the problems in this section. In nearly all the problems the difficult portion of the task is in the establishment of the original algebraic equations involved and not in their solution after they have been established. Only clear thinking and simple algebraic principles are involved in the solutions.

1. the professor's fish

Old Professor Warner boasted that his catch had been the best one made by any member of the Surf Club. He spread his hands a considerable distance apart to describe his prize catch.

"Well, how big *was* your fish?" one member asked him.

The professor paused a moment, then answered with a smile, "I remember the head measured nine inches. The tail was as long as the head and half the body, and the body was as long as the head and tail."

How long was the professor's prize fish?

2. floating down the river

Duncan was floating down the river on a raft when, half a mile downstream, his brother Turner took to the water in a canoe. Turner paddled on downstream at the best pace of which he was capable, then turned around and paddled back again—still at his best pace—and arrived at his starting point just as Duncan floated by.

Assuming Turner's best pace in still water is ten times that of the river current, what distance has he paddled?

3. one hundred dollars needed

Smith approaches his friends Jones, Brown, and Robinson and asks to borrow one hundred dollars.

Jones says, "You want $140 less than twice what I have."

Brown says, "You want $290 less than three times what I have."
Robinson says, "Any one of us will let you have the hundred, but I have just enough so that if you borrow all but $100 of my money and all but $100 of each of the others' money we will each one of us have exactly $100."
How much money does each have?

4. what is the product?

Divide a number by one more than itself. The answer is one-fifth.
Divide a second number by one more than itself. The answer is one-fifth the number.
Multiply these two numbers. What is the product?

5. joe nitwit

In his hurry to get on to more pleasant tasks Joe Nitwit turned on both intake pipes but forgot to close the drainpipe. The water tank was half full when his boss came by and, noticing the drainpipe open, closed it before going about his next most urgent task—firing Joe.
Assuming that it takes one of the intake pipes ten hours to fill the tank; the second intake pipe eight hours; and the drainpipe only six hours to empty the tank if both intake pipes are closed, how long did it take to fill the tank on this particular occasion?

6. the fly and the bicycles

Two cyclists, exactly a quarter of a mile apart, are approaching one another. One cyclist is proceeding at eight miles per hour and the other at twelve miles per hour.
A fly has been flying back and forth between the two cyclists at the constant speed of thirty miles per hour. When it reaches one bicycle it turns immediately and heads back toward the other. At the instant in question the fly has just left the wheel of the cyclist traveling at eight miles per hour and headed toward that of the cyclist traveling at twelve. It continues this backward-and-

forward motion between the onrushing cyclists until it meets a horrible death as they crash head-on.

How far did the fly travel during the interval in question?

7. birds in flight

Two boys were standing on a hill when a large flock of birds suddenly flew overhead. "Hey, look!" cried one. "Betcha there must be a hundred birds there."

The other boy was not only more sharp-sighted, but more calculative as well. "A hundred," he scoffed. "Nonsense! Why, there'd have to be again as many as there are and then half as many more than there are and then a quarter as many as there are—and even then, one more—before there would be a hundred."

How many birds were in the flock?

8. phil anthrope

Phil Anthrope, the eccentric millionaire, recently revisited his home town. As a memento of his visit he offered ten dollars to each of the boys in the town and six dollars to each of the girls. All the girls accepted his offer, but for some reason or other 40 per cent of the boys declined.

Assuming there was a total of 2240 boys and girls living in the town, how much did Phil give away?

9. the wire fence

A poultry farmer decided to build a wire fence along one straight side of his property. He planned to place the posts six feet apart, but after he bought the posts and the wire he found that he had miscalculated. He had five posts too few. However, he discovered that he could do with the posts he had by placing them eight feet apart.

How long was the side of the lot?

10. my favorite orange

My favorite orange weighs nine-tenths of its weight plus nine-tenths of a pound. What does it weigh?

11. candles

Recently installed in a garret room on the Left Bank, Painter was moodily staring out the window watching the darkness close over Paris at 9 P.M. Just then his one light bulb blinked feebly several times and went out, leaving him in almost total darkness. Luckily the concierge had left him a supply of two candles reserved for such emergencies, and he lit one immediately. An hour and a half later Painter decided that he needed more light and put the second candle into use, remembering that it was an inch shorter than the first one had been originally.

After two and a half hours more Painter suddenly noticed that the candles were of equal height. He mulled this fact over for an hour and a half before he concluded that perhaps candles burn at different rates. Just then, out winked the candle which had originally been shorter. A half-hour later the other one went out, leaving Painter in complete blackness to contemplate Paris and his mathematical problem.

How tall was each candle initially?

12. three drain pipes take——?

Consider a large tank which is receiving water at a constant rate from a large supply pipe and which is equipped with ten identical drain pipes. If the tank has Q gallons of water in it at time zero:

a. At the end of two and a half hours it will be empty if all ten outlets are open.

b. At the end of five and a half hours it will be empty if six of the outlets are open and the remaining four closed.

How long will it be before the tank will be empty if only three of the outlets are open and the remaining seven closed?

13. helen's age is?

A former Follies beauty, Helen Wentworth remained extremely chary of disclosing her correct age—especially to newspaper reporters. She instructed her husband to give roundabout answers whenever he might be asked for her age or for that of their daughter Louise.

Mr. Wentworth was no man to stretch the truth, but he did learn to disguise it neatly in this answer: "The combined ages of my wife Helen and our daughter Louise make an even hundred years. When Louise was half as old as Helen was when Helen was three times as old as Louise was when Louise was one-fourth as old as Helen was when Helen was twice Louise's age now, Helen was three times as old as Louise was when Louise was half as old as Helen was when Helen was as old as Louise will be four years after Louise will be half as old as Helen four years from now.

How old are Helen and Louise?

14. jack and jill

Charles Lutwidge Dodgson, the nineteenth-century mathematician, was better known under his pseudonym of Lewis Carroll. We are indebted to him for this fairly tricky problem.

Jack and Jill left home at three o'clock and walked along a level road, up a hill, back down the hill, and home—without stopping. They arrived home at nine o'clock. Their speed was four miles an hour on the level, three miles an hour uphill, and six miles an hour downhill.

How far did they walk and within half an hour what was the time they reached the top of the hill?

15. mrs. crabbe and the prunes

Mrs. Crabbe ran her boarding house fairly but strictly, with rules and regulations to be observed by everyone. Not knowing all the customs of the house, Mrs. Crabbe's new boarder helped himself to one more prune than was his share when the bowl started making the rounds. The dish next went to the Oldest Boarder, who quite inexcusably took two more prunes than his fair share. Then the Wag took three more than his share, just to see how Mrs. Crabbe would handle the untoward situation.

Her reaction was prompt; she seized the bowl, from which half the prunes were already gone, and portioned the remainder equally among her other boarders. Although each received one prune less than his due, none dared complain.

How many prunes would have been wasted if she had inverted the bowl over the Wag's head?

16. trust fund

Mr. Jenkins had just finished arranging a trust fund for his two children and his mind was awhirl with endless facts and figures.

"When do they start to receive the payments?" he was asked.

"Well now, let's see. I remember that the first payment comes due when the sum of their ages is forty-eight and when Peter is twice as old as he is now. He will then be ten years older than Betty was when Betty was twice as old as Peter was when Peter was one-third as old as Betty will be fifteen years from now."

How old are the children now?

17. what is the first time?

What is the first time after three o'clock when the minute hand will be as far past six as the hour hand is away from six?

18. round trip from abletown

At noon Jim, who is in training for a bicycle race, left Abletown to ride to Bakersville and back again. It is twenty-six miles each way. He did the double journey without stopping and maintained a uniform speed throughout.

Some time later Bob, trying out his new car, left Bakersville and drove—also maintaining a uniform speed—to Abletown and back again. Bob passed Jim on the latter's outward journey 7.5

miles from Bakersville and passed him again on his return journey 5.5 miles from Bakersville. Bob finished the double journey at 3:20 P.M.

What time was it when Jim was back at Abletown?

19. mr. wright and the insurance salesman

After a rough day at the office Mr. Wright was extremely short-tempered and in no mood for more questions. So when an insurance solicitor came around unexpectedly to see him that evening, his temper began to grow shorter every instant.

"And how old are your children, sir?" the agent asked him solicitously.

Mr. Wright fixed a withering eye upon him, then replied acidly, "The square of Archie's age added to the square of Barbara's age added to Chester's age equals the square of Chester's age. And when Archie is as old as Chester now, Chester will be four times as old as Archie now and twice as old as Barbara is now. Any more questions?"

How old are the three children?

20. philip gibson's children

Out walking with his three sons one Sunday afternoon, Philip Gibson ran into an old school friend he hadn't seen for many years.

"So you finally got married," his friend remarked, seeing the three boys.

"Sure thing, and I have these three children to prove it."

"How old are they now?"

Philip thought for a moment, then answered, "Well, the sum of their ages is twenty years. Jim's age added to Sam's is four times Tom's age, and Tom's age added to Jim's is four times Sam's."

How old are the three boys?

21. ducks

The other afternoon I asked my friend Frank, who enjoys raising "wild" ducks under conditions as near those which exist in their native habitat as practicable, "How many mallards do you have?" Somewhat to my astonishment his answer was, "If three-fourths of the ducks you can see could not be seen and half of those you cannot see could be seen, you would then fail to see a dozen dozen more ducks than you could see. However, if three-fourths of the ducks which you cannot see could be seen and half of those you can see could not be seen, you would then see a half-dozen dozen more ducks than you could not see."
How many mallards does Frank have?

22. ten-o'clock scholar

Junior was already much shrewder than his years suggested. As the proverbial "ten-o'clock scholar" he had to rely upon his wits once more as he scurried into the classroom after the morning bell had rung.
"Junior, you're late again. What time is it?" his teacher asked sternly.
He knew that he could easily steal back into her good favor by setting a problem for the rest of the class, so he answered. "Just add one-quarter of the time from midnight until now to half the time from now until midnight, and that's what time it is."
Doubletalk aside, what time was it?

23. packing crate

When Mrs. Rawlings got ready to pack her china for shipment to the family's new home in San Francisco, her husband designed a packing crate sixteen inches wider than deep and with a length five times the depth. But when he realized that he would not be able to get the crate through the front door, he had to take four inches off both length and width. He preserved the original capacity by increasing the depth one-quarter.
What were the original and the revised dimensions?

24. trains passing

A passenger train which is x times as fast as a freight train takes x times as long to pass when overtaking the freight train as it takes to pass when the two trains are going in opposite directions. What is the value of x?

25. wheat

Farmer Jones agreed to pay four hundred dollars in cash and a fixed number of bushels of wheat an acre as his rent.

When he signed the lease wheat was selling at a dollar a bushel and his rent amounted to eight dollars per acre. When the price of wheat rose to a dollar and a half per bushel, the farmer complained that his rent had been raised from eight dollars per acre to ten dollars per acre. How large was the farm?

26. the professor's window

I called on Professor Physics the other day and found him in a state of considerable agitation. It appeared that the celebration of a recent victory at football had taken the form of smashing his study window, though his annoyance at this was as nothing beside his indignation at the poor quality of the glass that had been used to repair it.

This, however, seemed to me to be a great deal of fuss about very little, for he told me that he had found that when the lower sash was half raised the total amount of light entering the room was 1 per cent less than when it was raised level with the upper sash.

"But surely," I said, "it is of very little consequence that half-opening the window stops 1 per cent more light than fully opening it?" "Oh, no, it doesn't," he retorted, "it stops 5 per cent more." "That's very odd," I replied. "I see that the sashes are equal in size and that the upper pane is rather darker than the lower; is it the same glass in both sashes?" But Physics was too much annoyed to answer this, and abruptly told me to find out for myself if I wanted to know.

Assuming that the framework is so narrow that it has no material effect upon the proportion of light admitted, what proportion of the light falling on it does each pane stop?

27. as the crow flies

"How far is it from Westburg to Southburg?"

"Do you mean as the crow flies?" asked Flyboy.

"Yes, as the crow flies."

"I really don't know," said Flyboy. "However, Westburg is fifty-six miles due west of Central City. Southburg is due south of Central City. I have flown the triangle formed by these three cities many times. I remember one trip in particular. During this trip I maintained a uniform speed at all times. It took me fifty minutes to fly from Central City to Southburg and an hour and a half to fly from Southburg back to Central City via Westburg."

"So what?" said Thickskull.

How far is it from Westburg to Southburg?

28. courier

A destroyer is directed to leave its position at the end of the fleet column and proceed to the head of the column, deliver a dispatch, and return to its former position. The destroyer and the column maintain a steady speed during the maneuver. When the destroyer regains its position Captain Fleck notes with interest that the end of the column now occupies the position which the head of the column had occupied when he first left his own position.

On the assumption that the time required for transfer of the dispatch, for the destroyer to get up to speed, or for the destroyer to reverse its direction is zero, what is the ratio of the destroyer's speed to that of the main column during this special mission? (Naturally, from a navigational point of view such an assumption is unrealistic, but—since it greatly simplifies the mathematics involved—it is made in the interest of this particular problem.)

29. outboard *vs.* speedboat

Seven equally spaced buoys mark the direct channel across Lake Euclid from Able to Baker landing. Outboard and Speedboat leave Able in their respective boats to go to Baker. Outboard goes direct. Speedboat has to stop at Charlie landing to get gas.

He notices that Outboard is just passing the third buoy as he arrives at Charlie and that Outboard is passing the fourth buoy as he leaves Charlie.

They arrive at Baker landing at the same time.

It is seven miles from Able to Baker and the channels from Able to Charlie and Baker to Charlie are at right angles to one another.

How far is it from Able to Charlie?

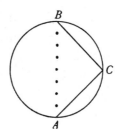

1. the professor's fish

Let h equal the length of the head of the fish. Then the length of the tail, t, is one-half the length of the body, b, plus the length of the head, or

$$t = h + \tfrac{1}{2}b$$

and the length of the body, which is equal to the length of the head and the tail, is
$$b = h + t.$$
Since $h = 9$ these equations become
$$t = 9 + \tfrac{1}{2}b$$
and $\qquad b = 9 + h = 9 + 9 + \tfrac{1}{2}b$
or $\qquad \tfrac{1}{2}b = 18 \qquad$ and $\qquad b = 36.$
Therefore $\qquad t = 9 + \tfrac{1}{2}(36) = 27$

and the entire length of the fish $= h + b + t = 9 + 36 + 27 = 72.$
The professor's prize fish was 6 feet long.

2. floating down the river

Let the speed of the current be x miles per hour. Turner's speed in still water is $10x$ miles per hour; downstream is $11x$ miles per hour and upstream is $9x$ miles per hour.

Obviously Duncan will float downstream at the rate of the current, which is x miles per hour, and it will take him $1/2x$ hours to float half a mile.

Let d equal the distance that Turner paddles each way. The round trip takes him $d/11x + d/9x$ hours, a sum that clearly equals $1/2x$. Therefore

$$\frac{d}{11x} + \frac{d}{9x} = \frac{1}{2x} \qquad \frac{d}{11} + \frac{d}{9} = \tfrac{1}{2} \qquad \text{or} \qquad 20d = {}^{99}\!/_2;$$

hence $d = {}^{99}\!/_{40}$ miles.

But d is the distance downstream. The total distance paddled is twice this amount, or $4^{19}\!/_{20}$ miles.

3. one hundred dollars needed

Jones' statement is equivalent to saying that $100 plus $140—or $240—will equal twice what he has. Jones has $120.

Brown's statement is equivalent to saying that $100 plus $290

—or $390—will equal three times what he has. Brown has $130.
Robinson must have $150; if he lends Smith $50, while Brown
lends him $30 and Jones lends him $20, each will have $100 left.
This may be stated algebraically as follows:

$$2J - 140 = 100 \quad \text{or} \quad J = 120$$
$$3B - 290 = 100 \quad \text{or} \quad B = 130$$
$$R + 120 + 130 - 300 = 100 \quad \text{or} \quad R = 150$$

4. what is the product?

Call the first number x. Then

$$\frac{x}{x+1} = \frac{1}{5}$$

Therefore $\qquad\qquad x = \frac{1}{4}.$

Call the second number y. Then

$$\frac{y}{y+1} = \frac{y}{5}$$

Therefore $\qquad y + 1 = 5 \qquad y = 4.$

$$xy = \text{one-quarter of four} = 1.$$

5. joe nitwit

With all three pipes open, water is going into the tank at the rate
of $\frac{1}{10}$ of a tank plus $\frac{1}{8}$ of a tank per hour and going out at the
rate of $\frac{1}{6}$ of a tank per hour for a net filling rate of

$$\frac{1}{10} + \frac{1}{8} - \frac{1}{6} = \frac{12 + 15 - 20}{120} = \frac{7}{120}$$

of a tank per hour. It would take $\frac{120}{7}$ hours to fill the tank under
these conditions and $\frac{60}{7} = 8\frac{4}{7}$ hours to fill it half full.

With only the two intake pipes open, no water is going out and
the net rate of filling is

$$\frac{1}{10} + \frac{1}{8} = \frac{8 + 10}{80} = \frac{9}{40}$$

of a tank per hour. It would take $4\frac{2}{7}$ hours to fill the tank under these conditions and $\frac{2\frac{4}{7}}{2} = 2\frac{2}{7}$ hours to fill it half full.

Under the conditions of the problem it will take

$$8\frac{4}{7} + 2\frac{2}{7} = 10^{5}\%_{3}$$

hours to fill the tank.

6. the fly and the bicycles

As in many mathematical problems, the idiotic behavior of the fly and the steadfast purpose of the two cyclists who maintain full speed up to the moment of collision are of no importance.

Nor are higher mathematics involved. True, the distance flown by the fly can be obtained by summing an infinite geometric series—but this is not necessary. Note that the cyclists are approaching at the rate of $8 + 12 = 20$ miles per hour and that it will take them $\frac{1}{80}$ of an hour to travel the quarter-mile which initially separated them. Since the fly is traveling at the rate of 30 miles an hour and

$$\text{rate } times \text{ time} = \text{distance}$$
$$30 \times \frac{1}{80} = \frac{3}{8}.$$

The fly traveled $\frac{3}{8}$ of a mile before the collision.

7. birds in flight

Let x equal the number of birds in the flock.

$$x + x + \frac{x}{2} + \frac{x}{4} + 1 = 100$$

or $2.75x = 99$ and $x = 36.$

Hence there were 36 birds in the flock.

8. phil anthrope

This may seem to be an impossible problem. However, if you offer $10 to each member of a certain group and 40 per cent decline, your average disbursement is $6 per member (each ten people in the group will be getting a total of $60, or $6 per individual). Since the girls received $6 each and the boys averaged $6 each, Phil Anthrope's giveaway must have averaged $6 per boy and girl in the town and, since the total population was 2240, he must have given away $13,440.

Expressed symbolically, if

$$B = \text{the number of boys, and}$$
$$G = \text{the number of girls,}$$
$$B + G = 2240$$

and $6B + \%_0(10G) = x$

where x is the total amount Phil gave away, or

$$6(B + G) = x.$$

But, as seen above, $B + G = 2240$; therefore

$$6(2240) = x$$
or $$x = \$13,440.$$

9. the wire fence

If $(x + 1) =$ the number of posts needed when 8 feet apart and $(x + 6) =$ the number of posts needed when 6 feet apart, it follows that (remembering that there must be one more post than there are intervals between posts) the length of the straight side is

$$8x = 6(x + 5)$$
or $$2x = 30$$
and $$x = 15.$$

The side was 120 feet long.

10. my favorite orange

The equation is

$$\%_0x + \%_0 = x$$
or $$9x + 9 = 10x$$
or $$x = 9.$$

My favorite orange weighs 9 pounds.

11. candles

If the taller candle is x inches high, its burning time of 6 hours makes its rate of consumption $x/6$ inches per hour. As the second candle is one inch shorter and burns only 4 hours, its rate of consumption is $(x - 1)/4$ inches per hour.

At 1 A.M. the longer candle has been burning 4 hours and the number of inches remaining is

$$x - \frac{x}{6}(4) = \frac{x}{3}.$$

At 1 A.M. the shorter candle has been burning 2½ hours and the number of inches remaining is

$$(x - 1) - \frac{(x - 1)}{4}(2\frac{1}{2}) = (x - 1)\frac{3}{8}.$$

But these two heights are equal, making

$$\frac{x}{3} = \frac{3(x - 1)}{8}$$

or $$8x = 9x - 9$$

or $$x = 9.$$

The candles were initially 9 and 8 inches long, respectively.

12. three drain pipes take——?

Let the capacity of the supply pipe be s gallons per hour and the capacity of any one of the drain pipes be d gallons per hour, then

1. $Q + 2\frac{1}{2}(s) = 2\frac{1}{2}(10d)$ from statement a
2. $Q + 5\frac{1}{2}(s) = 5\frac{1}{2}(6d)$ from statement b
3. $Q + x(s) = x(3d)$ where x is the number of hours

it will be before the tank is empty if only three of the drainpipes are open.

At first glance it would appear that these three equations with their four unknowns will not be sufficient to give the information desired, but let us see.

Subtracting the first equation from the other two gives

4. $3s = (33 - 25)d = 8d$ or $s = \dfrac{8d}{3}$ and

5. $(x - 2\frac{1}{2})s = (3x - 25)d.$

Now, substituting for s in the last equation its value, $8d/3$, gives

$$(x - 2\frac{1}{2})\frac{8d}{3} = (3x - 25)d$$

or, dividing by d and simplifying,

$$8x - 20 = 9x - 75,$$
or $$x = 55.$$

The tank will be empty at the end of 55 hours if three of the drain pipes are open and the remainder closed.

13. helen's age is?

Let x equal Louise's age and $x + a$ equal Helen's age. Since the sum of their ages is 100 years,

$$x + x + a = 100.$$

When Helen was twice Louise's present age, she was $2x$ years old. When Louise was one-fourth as old as Helen was when Helen was twice Louise's age now, she was $\frac{1}{4}(2x)$. When Helen was three times as old as Louise was when Louise was one-fourth as old as Helen was when Helen was twice Louise's age now, she was $3\{\frac{1}{4}(2x)\}$. When Louise was half as old as Helen was when Helen was three times as old as Louise was when Louise was one-fourth as old as Helen was when Helen was twice Louise's age now, she was $\frac{1}{2}[3\{\frac{1}{4}(2x)\}]$. Helen at this time would be a years older.

Helen's age can also be derived from the last portion of the problem statement, as follows: Helen will be $x + a + 4$ years old four years from now. When Louise is one-half as old as Helen will be four years from now, she will be $\frac{1}{2}(x + a + 4)$. When Louise was half as old as Helen was when Helen was as old as

Louise will be four years after Louise will be one-half as old as Helen will be four years from now, she will be

$$\tfrac{1}{2}\{4 + \tfrac{1}{2}(x + a + 4)\}.$$

When Helen is three times this old she will be

$$3[\tfrac{1}{2}\{4 + \tfrac{1}{2}(x + a + 4)\}].$$

Since these two expressions for Helen's age are equal, it follows that

$$a + \tfrac{1}{2}[3\{\tfrac{1}{4}(2x)\}] = 3[\tfrac{1}{2}\{4 + \tfrac{1}{2}(x + a + 4)\}]$$

or $\qquad a + \tfrac{3}{4}x = 6 + \tfrac{3}{4}(x + a + 4)$

or $\qquad a + \tfrac{3}{4}x = 6 + \tfrac{3}{4}x + \tfrac{3}{4}a + 3$

or $\qquad\qquad \tfrac{1}{4}a = 9 \qquad$ or $\qquad a = 36.$

Since $\qquad\qquad\qquad 2x + a = 100$

it follows that

$$2x + 36 = 100 \qquad \text{or} \qquad x = 32 = \text{Louise's age,}$$

and $\qquad\qquad x + a = 32 + 36 = 68 = \text{Helen's age.}$

Helen is sixty-eight and Louise is thirty-two.

14. jack and jill

Twenty-four miles and six-thirty.

This problem appears to be impossible to solve, but if you let the distance along the level road be d and the length of the hill h miles and proceed in a normal manner you get

$$\frac{d}{4} + \frac{h}{3} + \frac{h}{6} + \frac{d}{4} = 6$$

or $\qquad 3d + 4h + 2h + 3d = 6d + 6h = 72$

or $\qquad\qquad\qquad d + h = 12$

and $2(d + h)$ equals the total distance traveled, which is 24.

This same result would be obtained, of course, if you had noted that walking up a hill at 3 miles an hour and down the same hill at 6 miles an hour results in the average speed over this distance of 4 miles per hour. Since this is the same as Jack and Jill's speed on the level ground it follows that their average speed over the entire trip was 4 miles per hour and that the total distance walked must be 4 × 6 = 24 miles.

There is no way to calculate when they reached the top of the hill, but if there were no hill at all they would walk away from home until six o'clock and back the second three hours. If there were 12 full miles of hill they would spend four hours reaching the top and two hours coming home.

Thus they must have reached the top of the hill at some time after six and before seven, and six-thirty gives the time within half an hour.

15. mrs. crabbe and the prunes

To find the number of boarders it is not necessary to pay any attention to the provision that the first three took half the prunes. All we need to know is that since they got six more than their share the remaining boarders must have received six less and, since each remaining boarder lost one prune, there had to be six boarders left and a total of nine boarders.

To find the total number of prunes, set up the equation $3x + 6 = 6x - 6$ where x equals a fair share, and we have x equals 4. So a fair share was four; there were originally thirty-six prunes in the bowl.

Mrs. Crabbe would have dumped eighteen prunes over the Wag's head.

16. trust fund

Let P equal Peter's age now and B equal Betty's age now.

From the first statement it follows that $2P$ (Peter's age when he is twice as old as he is now) plus $[P + B]$ (Betty's age when Peter is twice as old as he is now) equals 48, or

$$2P + (P + B) = 48 \qquad \text{or} \qquad B = 48 - 3P.$$

Considering the second statement, when Peter is one-third as old as Betty will be fifteen years from now he will be

$$(B + 15) \, (\tfrac{1}{3})$$

and when Betty is twice this old she will be

$$2(B + 15) \, (\tfrac{1}{3})$$

and, since Peter will be twice his present age when he is ten years older than Betty at this point,

$$2P = 2(B + 15) \, (\tfrac{1}{3}) + 10.$$

Substituting the above value for B in this equation gives

$$2P = 2(48 - 3P + 15) \, (\tfrac{1}{3}) + 10$$

which reduces to

$$P = 13$$

therefore $\qquad B = 48 - 39 = 9.$

Peter is thirteen years old and Betty is nine years old at the present time.

17. what is the first time?

Let x equal the number of minutes past 3:30 when the minute hand is as far past six as the hour hand is short of it. It will then be $(30 + x)$ minutes past three o'clock.

The hour hand at this time will be, of course, under the conditions of the problem, x minutes short of six and will have moved since three o'clock a distance equal to $(15 - x)$ minutes.

Since the minute hand moves twelve times as fast as the hour hand, the distance it has moved since three o'clock must be twelve times the distance the hour hand has moved since three o'clock, or

$$(30 + x) = 12(15 - x)$$

or $\qquad 30 + x = 180 - 12x$

or $\qquad 13x = 150$

or $\qquad x = {}^{150}\!/_{13} = 11\tfrac{7}{13}.$

The first time after three o'clock the minute hand will be as far past six as the hour hand is short of six is 3:41$\tfrac{7}{13}$.

18. round trip from abletown

Between their two meetings Jim covers 7½ miles into Bakersville and 5½ miles back toward Abletown, a total of 13 miles, at a uniform speed of s miles per hour. During the same interval of time Bob covers 18½ miles into Abletown ($26 - 7\frac{1}{2} = 18\frac{1}{2}$) and 20½ miles back toward Bakersville ($26 - 5\frac{1}{2} = 20\frac{1}{2}$), a total of 39 miles, at a uniform speed of r miles per hour. Since these two times are equal

$$\frac{13}{s} = \frac{39}{r}$$

or $\qquad\qquad\qquad r = 3s.$

At the time of their second meeting Jim had been pedaling for $(26 + 5\frac{1}{2})/s = 31\frac{1}{2}/s$ hours. Also, the time from this meeting until Bob returns to Bakersville at 3:20 P.M. is $5\frac{1}{2}/r = 5\frac{1}{2}/3s$ hours. Since the sum of these two times equals the difference between the time of Bob's return to Bakersville and Jim's departure from Abletown it follows that

$$\frac{31\frac{1}{2}}{s} + \frac{5\frac{1}{2}}{3s} = 3\frac{1}{3} = 1\frac{0}{3}$$

or $\qquad\qquad 10s = 94\frac{1}{2} + 5\frac{1}{2} = 100$

and $\qquad\qquad\quad s = 10.$

Jim's speed is 10 miles per hour, and he was back at Abletown at 5:12 P.M.

19. mr. wright and the insurance salesman

Let the ages of the children be represented by their initials—a, b, and c. From the statement that the sum of the square of Archie's age, the square of Barbara's age, and Chester's age equals the square of Chester's age, it follows that

$$a^2 + b^2 + c = c^2.$$

Since, when Archie is as old as Chester is now, Chester's age will be $c + (c - a)$, from the second statement of the problem you get

$$c + (c - a) = 4a \quad \text{or} \quad a = \tfrac{2}{5}c$$
and $$c + (c - a) = 2b \quad \text{or} \quad 2b = 2c - a.$$

Substituting for a its value $\tfrac{2}{5}c$, the last equation becomes

$$2b = 2c - \tfrac{2}{5}c = \tfrac{8}{5}c \quad \text{or} \quad b = \tfrac{4}{5}c.$$

Substituting these values of a and b in the first equation gives

$$(\tfrac{2}{5}c)^2 + (\tfrac{4}{5}c)^2 + c = c^2$$
or $$4c^2 + 16c^2 + 25c = 25c^2$$
or $$5c^2 = 25c$$
or $$c = 5$$
and so $$a = 2 \quad \text{and} \quad b = 4.$$

Archie is two, Barbara is four, and Chester is five.

20. philip gibson's children

Let the ages of the children be represented by their initials—J, S, and T. The conditions of the problem can then be expressed as

$$J + S + T = 20$$
$$J + S = 4T$$
$$J + T = 4S.$$

Subtracting the second equation from the first gives

$$T = 20 - 4T \quad \text{or} \quad T = 4.$$

Subtracting the third equation from the first gives

$$S = 20 - 4S \quad \text{or} \quad S = 4$$

(it is obvious from the symmetry of the above equations that S would prove to be equal to T). Substituting these values in the first equation gives

$$J + 4 + 4 = 20$$
or $$J = 12.$$

Jim is twelve and Sam and Tom are both four.

21. ducks

Let x equal the number of ducks that you can see and
y equal the number of ducks that you cannot see.
From the first statement the number you can see under the supposed situation is 12 dozen less than the number you cannot see, or

$$\tfrac{1}{4}x + \tfrac{1}{2}y = \tfrac{3}{4}x + \tfrac{1}{2}y - 12 \times 12$$

or $\quad \tfrac{1}{2}x = 12 \times 12 \quad$ or $\quad x = 24 \times 12 = 24$ dozen.

From the second statement the number you can see is 6 dozen more than the number you cannot see, or

$$\tfrac{1}{2}x + \tfrac{3}{4}y = \tfrac{1}{2}x + \tfrac{1}{4}y + 6 \times 12$$

or $\quad \tfrac{1}{2}y = 6 \times 12 \quad$ or $\quad y = 12 \times 12 = 12$ dozen
and $\qquad\qquad\qquad x + y = 36$ dozen.

Frank has 36 dozen mallards.

22. ten-o'clock scholar

Let the present time be t. One-quarter of the time from midnight until now is $t/4$ and one-half the time from now until midnight is $(24 - t)/2$, or

$$\frac{t}{4} + \frac{(24 - t)}{2} = t.$$

Multiplying both sides by 4 and collecting terms gives

$$5t = 48 \quad \text{or} \quad t = 9.6 \text{ hours.}$$

The time was 9:36 A.M.

23. packing crate

Let the original depth be d. Then the original dimensions of the crate were $(d + 16) \times d \times 5d$, and the modified dimensions would be $(d - 4 + 16) \times \tfrac{3}{4}d \times (5d - 4)$. Since both designs had the same volume, it follows that

$$(d + 16) \times d \times 5d = (d + 12) \times \frac{5d}{4} \times (5d - 4)$$

or $\qquad 5d^3 + 80d^2 = 2\frac{3}{4}d^3 + 70d^2 - 60d$

or $\qquad \frac{3}{4}d^3 - 10d^2 - 60d = 0$

or $\qquad d^2 - 8d - 48 = 0$

or $\qquad (d - 12)(d + 4) = 0$

or $\qquad d = 12 \text{ or } -4.$

Since minus 4 has no significance in such a problem, d must have been equal to 12 inches and the crate, which *originally* was $28 \times 12 \times 60$ had revised dimensions of $24 \times 15 \times 56$.

24. trains passing

If d equals the combined length of the two trains and v equals the speed of the freight

$$t_1 = \frac{d}{xv - v} = \quad \text{time it takes the passenger train to pass when overtaking, and}$$

$$t_2 = \frac{d}{xv + v} = \quad \text{time it takes the passenger train to pass when the trains are going in opposite directions}$$

Since $\qquad\qquad\qquad t_1 = xt_2$

$$\frac{d}{v(x - 1)} = \frac{xd}{v(x + 1)}$$

or $\qquad\qquad (x + 1) = x(x - 1)$

giving $\qquad\qquad x^2 - 2x - 1 = 0$

or $\qquad\qquad x = 1 \pm \sqrt{2}$

and, keeping only the positive value, it follows that $x = 1 + \sqrt{2}$.

25. wheat

Here is a straightforward algebraic solution:

Let x equal the number of bushels of wheat the farmer pays per acre and y the number of acres in the farm. It follows that the rent the farmer was paying per acre was $400/y$ plus the value of x bushels of wheat. Since this amount was $8 when wheat was $1 a bushel and $10 when wheat was $1.50 a bushel, it follows that

and
$$\frac{400}{y} + x(1) = 8$$

$$\frac{400}{y} + x(1.50) = 10.$$

Subtracting the first equation from the second,

$$x(0.50) = 2$$
or
$$x = 4,$$

and substituting this value of x in the first equation gives

$$\frac{400}{y} + 4(1) = 8$$

or
$$4y = 400$$
or
$$y = 100,$$

so the farm consisted of 100 acres.

This solution involves slightly more direct reasoning:

The farmer's rent in wheat was obviously 4 bushels an acre (a rise of 50 cents per bushel resulted in a rental increase of $2).

He also must have been paying $4 per acre in cash (when wheat was $1 a bushel 4 bushels were worth $4 and, since the total rent was then $8, the cash portion thereof must have been $4).

Since his total cash outlay was $400 there must have been 100 acres in the farm.

26. the professor's window

A piece of glass allows a certain fraction of the incident light to pass through it. Consequently, where light passes through two pieces of glass in turn, the fraction passed is the product of the fractions passed by the two pieces separately.

Let the fraction of light passed by the upper pane be x and by

the lower pane be y. Where the panes overlap the fraction passed will be, therefore, xy. With the window one-half open the top one-fourth of the total area (where there is only the upper pane of glass) passes an xth part of the incident light, or $x/4$ of the total light falling on the window. The next fourth of the total area (where both pieces of glass overlap) passes an xyth part of the incident light, or $xy/4$ of the total light falling on the window. The next fourth of the total area (where there is only the lower pane of glass) passes a yth part of the incident light, or $y/4$ of the total light falling on the window. The lower fourth of the total area (where there is no glass) passes all of the incident light, or one-fourth of the total light falling on the window. Adding these quantities together gives as the total fraction of light transmitted through the half-open window

$$A = \frac{(x + xy + y + 1)}{4}.$$

The fraction stopped by the window under these conditions is

$$a = 1 - \frac{(x + xy + y + 1)}{4} = \frac{(3 - x - xy - y)}{4}.$$

By similar reasoning it follows that the fraction of light transmitted when the window is wide open is

$$B = \frac{(2xy + 2)}{4} = \frac{(xy + 1)}{2}$$

and the fraction stopped under these conditions is

$$b = 1 - \frac{(xy + 1)}{2} = \frac{1 - xy}{2}.$$

But $\qquad A = 0.99B \qquad$ and $\qquad a = 1.05b$

hence $\qquad \dfrac{(x + xy + y + 1)}{4} = \dfrac{99(xy + 1)}{200}$

and $\qquad \dfrac{(3 - x - xy - y)}{4} = \dfrac{105(1 - xy)}{200}.$

Adding these two equations together gives

$$1 = \frac{99xy + 99 + 105 - 105xy}{200} = \frac{204 - 6xy}{200}$$

or $xy = \frac{2}{3}.$

Substituting this value in the first equation gives

$$\frac{(x + \frac{2}{3} + y + 1)}{4} = \frac{99(\frac{2}{3} + 1)}{200} = {}^{165}\!/_{200} = {}^{33}\!/_{40}$$

or $x + y = {}^{33}\!/_{10} - \frac{2}{3} = {}^{49}\!/_{30}$

or $x + \dfrac{2}{3x} = {}^{49}\!/_{30}$

or $30x^2 + 20 = 49x$

or $x = {}^{5}\!/_{6} \text{ or } {}^{4}\!/_{5}$

$$y = {}^{4}\!/_{5} \text{ or } {}^{5}\!/_{6}.$$

The darker pane stops one-fifth and the lighter pane one-sixth of the light falling on it.

27. as the crow flies

Let r be Flyboy's speed in miles per minute and x be the distance in miles from Central City to Southburg. The distance from Southburg to Westburg is then, by the Pythagorean Theorem,

$$\sqrt{x^2 + 56^2}$$

and, since rate times time equals distance,

$$50r = x \qquad \text{or} \qquad r = \frac{x}{50}$$

and $90r = 56 + \sqrt{x^2 + 56^2}$

or, upon replacing r by its equivalent $x/50$,

$$\frac{90x}{50} - 56 = \sqrt{x^2 + 56^2}.$$

Squaring both sides gives

$$\frac{81x^2}{25} - \frac{18(56)x}{5} + 56^2 = x^2 + 56^2$$

or

$$\frac{56x^2}{25} = \frac{18(56)x}{5}$$

or

$$\frac{x}{5} = 18$$

or $x = 90$ and $\sqrt{56^2 + 90^2} = \sqrt{11236} = 106$

It is 106 miles from Westburg to Southburg.

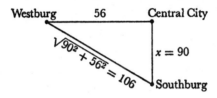

28. courier

Let L be the length of the main column and x the distance the main column travels while the courier goes from the rear of the column to the head of the column. Then

$L - x$ equals the distance the column travels while the courier goes from the head of the column back to the rear,

$L + x$ equals the distance the courier travels to get to the head of the column, and

$L - (L - x) = x$ equals the distance the courier travels to get from the head of the column back to the rear.

Since both courier and column maintain constant rates of speed the ratio of these distances must be equal, or

$$\frac{x}{L + x} = \frac{L - x}{x}$$

or $$x^2 = L^2 - x^2$$

or $$x = \frac{L\sqrt{2}}{2}$$

The distances traveled are

by courier on its way forward	$L(1 + \frac{1}{2}\sqrt{2})$
by courier on its way back	$L(\frac{1}{2}\sqrt{2})$
by courier both directions	$L(1 + \sqrt{2})$
by column total distance	$L.$

From which it follows that the ratio of the courier's speed to that of the column, which must equal the ratio of these distances, is $L(1 + \sqrt{2})/L = (1 + \sqrt{2}) = 2.414$ approximately.

29. outboard *vs.* speedboat

Since the two channels to Charlie are at right angles to one another, Speedboat's journey was along the two sides of a right-angle triangle, while Outboard went along the hypotenuse. Thus we have the famous Pythagorean Theorem of $a^2 + b^2 = c^2$, where a is the distance from Able to Charlie, b is the distance from Charlie to Baker, and c is the distance across the main channel from Able to Baker.

Now for something a little tricky. The seven equally spaced buoys divide the channel from Able to Baker into eighths (not sevenths). The first buoy is one-eighth of the way; the second two-eighths or one-quarter; the third three-eighths, the fourth half-way, and so on.

Speedboat went three units of distance while Outboard was on his way to the third buoy; four more while Outboard was covering the distance from the fourth buoy to Baker.

Let $3x =$ distance from Able to Charlie.

Then $4x$ must equal the distance from Charlie to Baker.

In our equation $a^2 + b^2 = c^2$ a becomes $3x$, b becomes $4x$, and $c = 7$.

$$9x^2 + 16x^2 = 49$$
$$25x^2 = 49$$
$$x = \frac{7}{5}$$
$$3x = \frac{21}{5} = 4\frac{1}{5} \text{ miles}$$

It is $4\frac{1}{5}$ miles from Able to Charlie.

the odds: 3
explorations in probability

The probable and the possible have always fascinated human beings. The long chance, and the possibility of winning it, remains in one way or another a part-time pursuit of mankind. A great many probabilities in life cannot be gauged, but many others can. On those that can be measured, mathematics brings to bear a scientific exactness that often seems uncanny to the uninitiated.

The probability of an event happening can be measured if we can determine the number of equally likely ways in which it can occur (*favorable cases*) and the number of equally likely ways in which it cannot occur (*unfavorable cases*).

Suppose we throw two normal six-sided dice on the table, one red and one green. There are six ways the red die can come up, and for each one of these six ways there are six ways the green die can come up. There are therefore thirty-six ways the dice can fall on the table. If the two dice are tossed at random, any of these thirty-six ways is equally likely; before a particular toss we say that there is a chance of one in thirty-six (a probability of $\frac{1}{36} = 0.02777 - - -$) that any particular combination will come up. Let us define a total of 4 as success and any other total as failure. There are three different ways of obtaining a total of 4: 1 on the red die and a 3 on the green; a 2 on each die; and a 3 on the red die and a 1 on the green. There are thus three favorable and thirty-three unfavorable cases out of a total of thirty-six possibilities. The probability of success (total of 4) is the number of favorable cases divided by the number of possible cases ($\frac{3}{36} = \frac{1}{12} = 0.08333$ – – –), and the probability of failure (any other total) is the number of unfavorable cases divided by the number of possible cases ($\frac{33}{36} = \frac{11}{12} = 0.91666$ – – –).

While this example was a specific case, the principle is quite general: *The probability that an event will occur equals the number of cases favorable to success divided by the total number of possible cases.* For this definition of probability to hold, however, it is essential that each of the possible cases be equally likely. From this definition it follows that a probability of one indicates certainty and a probability of zero indicates impossibility.

Take another example. Suppose we are asked to determine the probability that a single card drawn at random from a well-shuffled standard pack of playing cards will be either an ace or a king. The number of favorable cases is clearly eight, since there are four aces and four kings in the pack; the total number of cases is fifty-two, the number of cards in the pack. The desired probability is therefore $\frac{8}{52} = \frac{2}{13} = 0.1538$.

For simple events, as in the example of the cards, we can compute the number of cases—and thus the probability of success—almost by inspection. For compound events (a series of draws from a deck or repeated tosses of a coin) the problem is more complicated but can be simplified by dividing it into one of three categories, depending upon whether or not the events in which we are interested are independent, exclusive, or related.

1. Two or more events are said to be *independent* if the probability of any given event occurring is not affected by the result of the other events. If we toss a coin repeatedly, each toss is independent of each other toss; so are repeated draws from a pack of cards if the card drawn is replaced and the pack reshuffled each time. It is a simple matter, then, to determine that the probability of the occurrence of a series of independent events in a speci-

fied order is simply the product of the individual probabilities.

For example, assume that you are tossing a perfect coin which has a probability of one half that it will come up heads and one half that it will come up tails. The probability of getting two heads and two tails in the order HHTT is the product of the individual probabilities:

½ times ½ times ½ times ½ = ¹⁄₁₆.

Or, tossing a normal die repeatedly, the probability that you will throw a 6, followed by a 5, followed by an odd number, followed by an even number, is the product of their individual probabilities:

⅙ times ⅙ times ½ times ½ = ¹⁄₁₄₄.

2. Two or more trials are said to be *mutually exclusive* when they are related in such a manner that only one of them can possibly happen. In this situation the probability that any one of them will occur is the sum of their individual probabilities. Thus the probability of rolling a six with one die is clearly ⅙. So is the chance of rolling a five. Since these are mutually exclusive events (if one of them has occurred, the other could not possibly have occurred), the probability of rolling either a five or a six on any given toss is

⅙ + ⅙ = ²⁄₆ = ⅓.

Similarly, the probability of drawing the ace of spades, the king of hearts, the queen of diamonds, or the jack of clubs is

¹⁄₅₂ + ¹⁄₅₂ + ¹⁄₅₂ + ¹⁄₅₂ = ⁴⁄₅₂ = ¹⁄₁₃.

3. Two or more events are said to be *related* if the probability of a particular event occurring on any given trial is dependent upon the outcome of one or more of the previous trials. Probabilities of related events are quite difficult to compute at times. However, by restating the problem you can usually convert it into one involving independent or mutually exclusive events (or a combination of independent and mutually exclusive events) and thus reduce the problem to one for which the solution is known.

Here is a simple example: What is the probability of drawing a heart from a well-shuffled standard pack of playing cards and then, without replacing the card drawn first, drawing an ace or a king ·from the same pack? This problem can be restated as follows:· What is the probability of drawing the ace or king of hearts on the first draw and some other ace or king on the second draw (without replacement), or drawing some heart other than the ace or king on the first draw and any ace or king on the second draw (still without replacement)? The restatement establishes two mutually exclusive events which can be handled in this fashion:

The probability of drawing the ace or king of hearts on the first draw is $\frac{2}{52}$. Having drawn the ace or king of hearts, there are fifty-one cards left—of which seven are aces or kings. Therefore, the probability of drawing an ace or king on the second draw is $\frac{7}{51}$ and the probability of both these events occurring in the order named—since they are independent—is their product:

$$\frac{2}{52} \times \frac{7}{51} = \frac{14}{2652}.$$

Further, the probability of drawing a heart other than the ace or king on the first draw is $\frac{11}{52}$. Having drawn a heart other than the ace or king, there are fifty-one cards left—of which eight are aces and kings. Therefore the probability of drawing an ace or king on the second draw is $\frac{8}{51}$ and the probability of both these events occurring in the order named is also their product:

$$\frac{11}{52} \times \frac{8}{51} = \frac{88}{2652}.$$

These two series of events obviously represent the only way in which a heart can be drawn on the first draw and (without replacement) an ace or a king on the second draw; the probability that one or the other will occur becomes the sum of their individual probabilities as previously computed:

$$\frac{14}{2652} + \frac{88}{2652} = \frac{102}{2652} = \frac{1}{26}.$$

PERMUTATIONS AND COMBINATIONS

We have just seen that the computation of a complex problem in probability can frequently be expressed in terms of simple probabilities and that these simple probabilities can usually be expressed as a ratio of the number of favorable cases to the total number of possible cases. This is not always as easy as it sounds. The calculation of the number of cases, in fact, can prove to be quite difficult; for practical purposes they are frequently impossible to determine except by use of the basic formulas of permutations and combinations.

Since these formulas are somewhat involved, they are not given here; the reader who wishes to explore this area can find their derivation and some examples in any elementary textbook on probability.

The problems in this section (with the exception of numbers 9 and 14) can be solved without going into operations more complex than calculating simple probability or—at most—figuring out how to restate the problem.

1. cue's chance

"Take you on at billiards," said Cue to Ball.

"Right you are," said Ball. "We'll play five games."

"I ought to beat you, Ball, you know," said Cue.

"Of course," said Ball. "You'll have to give me odds."

"How do my chances of winning compare with yours?" asked Cue.

"Well," replied Ball, "I'll tell you. Your chance of winning three of the five games and your chance of winning four of the five games are, according to my calculations, exactly the same."

What is the chance (according to Ball's calculations) that Cue will win all five games?

2. transfers

Professor Knowsodds had five yellow balls in bag 1. He transferred one of them to bag 2, which contained an unknown number of green balls. After shaking up bag 2 he selected a ball at random and, without looking at it, transferred it to bag 1. He now thoroughly mixed the balls in bag 1 and selecting a ball from it at random, again without looking at it, transferred it to bag 2.

The professor then told his class that if he were to select a ball at random from bag 2 the chances it would be a green ball were three to two. He then asked them "How many green balls were there originally in bag 2?"

3. the odds are two to one

A pack of cards is divided by cutting it into two unequal portions.

If a card is drawn from portion A, the odds are two to one against its being a red card.

A red card is next transferred from portion B to portion A. Now the odds are two to one against a card drawn at random from portion B being black.

How was the pack originally divided?

4. how many left?

Professor Askem was conducting a demonstration before his probability class. "These two bags," he told the class, "have similar contents. Each bag contains four blue marbles, four red marbles and four yellow marbles. I will now select at random from Bag Number 1 enough marbles (but only just enough) to ensure that my selection includes at least two marbles of any one color, plus one marble at least of either of the other colors, and transfer them to Bag Number 2.

"Now," said the professor, again closing his eyes, "I transfer from Bag Number 2 back to Bag Number 1 enough (but only just enough) marbles to ensure that there will be at least three marbles of each of the three colors in Bag Number 1."

How many marbles did the professor leave in Bag Number 2?

5. tennis at lower dropshot

Eighty-two players entered the Lower Dropshot Club tennis championship this year. Eighteen first-round matches reduced the field to sixty-four, whereupon the regular elimination followed. How many matches were played altogether?

6. face cards

Green, Pink, and White were waiting for a fourth at bridge. Green took an ordinary pack of playing cards and divided it into three portions. Of these three, White's portion was the smallest.

Green looked at his cards. "Here's an interesting fact," he said. "If my cards were shuffled and two cards drawn at random, it's an even-money chance that neither would be a face card [a king, queen, or jack]."

"The same is true of my cards," said Pink. "I think that is even more interesting."

"It is indeed," said Green. He then turned to White and asked, "Can you tell us anything interesting about your cards?"

Could he?

7. the king of hearts

Two identical packs of cards are shuffled thoroughly. One card is drawn from pack A and shuffled with pack B.

The top card of pack A is turned up. If this card is the queen of hearts, what is the chance that the top card of pack B will be the king of hearts?

8. birthdays

Forty doctors were present at a meeting of the County Medical Association. At one of the subcommittee meetings Dr. Pinkpill remarked, "This is my birthday." This remark led to a discussion of birthdays in general and it turned out that Dr. Sawbones, the surgeon from Podunk, and Dr. Fixem of Siwash were both born on the same day of the year.

As there were only six doctors on the subcommittee, this seemed to the assembled doctors to be a most unusual event. One of them wrote to his friend, Professor Knowsodds, to find out just what the chances were.

Neglecting the effect of leap year (that is, assuming each year has 365 days), what is the chance that, of six people selected at random, two will have the same birthday?

9. the walk to 15th and m

The thriving metropolis of Podunk is laid out in squares. A man who lives at 8th Street and Avenue F bowls at alleys located at 15th Street (which is seven blocks north) and Avenue M (which is seven blocks east).

It is a nice fourteen-block walk, so every time he bowls he walks back and forth. Since all streets are equally desirable for walking he chooses his route at random except that he plans to use every possible route before repeating any particular one.

How many possible routes are there?

10. the three prisoners

Three criminals are convicted of participating in the same crime. The penalty is death for the one who bears the greatest degree of guilt.

They are in prison waiting for the judges to decide which one is the most guilty and prisoner A asks a guard if he can give him any information. The guard replies, "All I can tell you is that B is not going to be executed and that they have not considered your case at all."

What is the chance that A will be executed?

11. the dictator

The new dictator devised an unusual punishment for his five most important prisoners. He announced that two would be shot and three would go free. He further announced that, while he had decided which two were the unlucky ones, none of the five would know their fate until execution day.

On that day all five were to walk through the gate to the prison yard in any order they chose. If the first two out were the ones to be shot the gate would shut and the last three would go free. Otherwise all five would march out and line up before the far wall for the execution of the two victims.

On the fateful day A and B go through the gate first and it does not close. A is something of a mathematician, so as he marches across the courtyard to the wall he computes his chance of survival.

What is it?

12. more tennis at lower dropshot

As an aftermath of the tournament at Lower Dropshot, Cut and Smash agree to play a series of matches. They also agree to toss a coin for first serve in each match.

Someone asks Professor Knowsodds, who is going to referee the series, about the number of matches. His reply: "It is exactly even money that Cut will win at least three of the tosses."

How many matches had they agreed to play?

13. cubes

Professor Probability gave each of his sons a large wooden cube. Each cube had each of its six faces painted a bright cheerful color.

Said the professor to his sons:

"I want each of you to assign to each face of his cube one of the numbers 1 to 6, using each number once and only once. Then write this number in each corner of the face assigned to it.

"Next, take your saws and cut each cube into eight smaller ones. Each of these cubes will then have three numbered and three unnumbered faces. The sum of the numbers on its three numbered faces we will call its *total*."

The sons did as they were told and in a short time each of them produced for inspection his eight smaller cubes.

"Good," said Professor Probability. He examined the cubes. "I now set aside two of these; one has a *total* of nine and one a *total* of seven. I put the remainder of the cubes into a hat. If I draw one out at random the chances are two to one against it having a total of nine."

"How many cubes in the hat have a *total* of fourteen?"

On the assumption that the professor has the least number of sons for which the problem has a solution, what is the answer to his question?

14. king's town

Twenty-four players competed in the recent chess tournament at King's Town. The committee divided them into two sections. In each section each competitor played one game against every other player.

There were sixty-nine more games in Section B than in Section A.

Gambit, unbeaten in Section A, scored 5½ points. (Win = 1 point; draw = ½ point.)

How many of Gambit's games were draws?

15. professor of ancient history

As the professors were taking their seats on the platform a freshman asked "Who is the man with the white beard?"

"That's the Professor of Ancient History," answered his right-hand neighbor.

"So it is," said the next man. "Ugly old boy, isn't he?"

Assuming that one of these speakers makes a point of telling the truth three times out of four and the other tells it four times out of five, what is the chance that the white beard belongs to the Professor of Ancient History?

16. two spades

A bridge hand contains seven spades among its thirteen cards. What is the chance that two cards drawn at random will both be spades?

17. smithley's problem

During Smithley's last visit to the South Sea Island of Ha he happened to be there on 8–8 Day. This is their main holiday—the eighth day of the eighth month—and on this day all the natives wear a special scarf which is symbolic of their race. For this one day, then, Smithley had no trouble identifying the race to which his friends belonged, but he quickly found that this did not eliminate his troubles.

He walked up to a group of four natives who were all wearing green scarfs (and hence members of the Green race—that proud race who tell the truth one time in four when asked a series of questions and who, when in a group and conversing, answer truthfully but completely at random, one time in four) just in time to hear one of them (Ah) make a statement in reply to a question. Turning to one of the others (Bah) he asked him, "Did he speak the truth?" Bah answered something which Smithley could not understand, so Smithley turned to one of the two remaining natives (Cah) and asked him, "What did Bah answer?" Cah answered the question, but once more Smithley could not understand the reply. So he turned to the last native, who spoke English, and asked him, "Dah, what did Cah say?" To his astonishment Dah replied:

"I, Dah, assert,
That Cah denied,
That Bah said,
That Ah lied!"

Given that Ah, Bah, Cah, and Dah each speak the truth once in four times (that is, the *a priori* probability that any one of them is telling the truth is ¼) what is the probability that Ah told the truth when he made his original statement?

18. the odds against smith

Smith, Brown, and Jones were so far superior to the other members of the club that they were the only entrants in the championship matches. So it was decided that one entrant (to be chosen by lot) would receive a first round bye and that this entrant would meet in the finals the winner of a semi-final match between the other two.

Even so, Smith's chances are not very bright. While Brown and Jones are equally strong players, the odds are two to one against Smith in a match against either of them.

What is the probability that Smith will win the championship?

19. professor knowsodds' reply

Discussion at the club had turned to one of those puzzles in which each of four men goes home with someone else's hat. "Suppose there had been five of these absent-minded professors," said someone. "In how many ways could each of them be given the wrong hat?"

"I've seen the answer somewhere," said Misinformed. "I think it is fifty." "I don't," said Opinionated. "I'm pretty sure its forty-five." "And I beg to differ from both of you," said Knowsall. "I forget how it's worked out, but I've been told that it is forty-two." He turned to Professor Knowsodds. "That's right, isn't it, Professor? You know all about these things."

"Actually," said the professor, "I've forgotten what the answer is. But you do not have to know the answer to realize that your three guesses are, on the face of them, wrong."

On what did Professor Knowsodds base his statement?

20. opinionated is not satisfied

Opinionated was not satisfied with Professor Knowsodds' reply in the preceding problem as to why 45 was not the proper answer to the question "In how many ways could each of five absent-minded professors be given the wrong hat?" and asked him to supply the answer.

How many ways are there?

21. a pair of dice

Joe has just rolled a pair of dice and is looking at them.

What are the chances that he has rolled a pair of sixes under each of these circumstances:

1. Joe announces "I did not roll any fours."
2. Bill asks Joe "Did you roll at least one six?" and Joe replies "Yes."
3. While one die is still rolling, Joe announces "I have at least one six."
4. Joe announces "I rolled at least one six."

22. girls should live in brooklyn

A hard-working young businessman used to spend his evenings calling on one of two girls. The young man must have had something to offer because both girls would wait for him each night. He would leave his office at a random time around eight o'clock and go to the nearby subway station and take the first express train to come along. Both Brooklyn and Van Cortlandt trains stopped at the same platform and, since one girl lived near Van Cortlandt Park and the other in Brooklyn, he thought his choice of which girl to see was a random one.

To his surprise he discovered over a period of time he was seeing the girl who lived in Brooklyn about three times as often as the other girl. After a check with the subway system confirmed his belief that there was an equal number of express trains in each direction at that time of day he was at a loss to account for this fact. Can you?

23. the three chess players

Three chess players agree to play a series of games for a prize. The prize will go to the first player who can win two games in succession. The players draw lots to see which two shall play the first game. From then on the winner plays against the man who has just sat out. Assuming the three players to be of exactly equal skill, what are their respective chances?

24. the five balls

Professor Probability placed a box on his desk and requested members of the class (who were told nothing about the number or color distribution of the balls in the box) to draw out five balls selected at random. All five were blue.

"That that would happen," said the professor, "was exactly an even-money chance."

What is the smallest number of balls the professor could have placed in the box and how many of them were blue?

25. how many balls?

"This bag contains an assortment of blue and red balls," said Professor Probability. "If I draw two balls from it at random, the probability that I will have drawn two red balls is five times the probability that I will have drawn two blue balls. Furthermore, the probability that I will have drawn one ball of each color is six times the probability that I will have drawn two blue balls."

Turning to his smartest student, the professor continued, "Wesley, how many red and how many blue balls are there in the bag?"

solutions

1. cue's chance

Let Cue's chance of winning any one game be p and his chance of losing any one game be q. Then

$$p + q = 1.$$

As discussed in the introduction to this section, there are ten possible ways in which Cue can win three games and lose two, each of which has a probability of p^3q^2. (As a matter of possible interest, these ten ways are: WWWLL, WWLLW, WLLWW, LLWWW, WWLWL, WLWLW, LWLWW, WLWWL, LWWLW, and LWWWL.) The total probability that Cue will win three games and lose two is, therefore, $10p^3q^2$.

Similarly, there are five ways in which Cue can win four games and lose one (WWWWL, WWWLW, WWLWW, WLWWW, and LWWWW), each with a probability of p^4q, for a total probability of $5p^4q$.

According to Ball's calculations these two probabilities are equal, or

$$10p^3q^2 = 5p^4q$$

or

$$2q = p$$

but

$$p + q = 1.$$

Substituting for p its equivalent $2q$ gives

$$2q + q = 1 \quad \text{or} \quad q = \frac{1}{3}$$

and

$$p = \frac{2}{3}.$$

If Cue's chance of winning any single game is $\frac{2}{3}$, his chance of winning all five games is $(\frac{2}{3})^5 = \frac{32}{243}$.

2. transfers

Originally there were five yellow balls in bag 1 and x green balls in bag 2. After the first transfer there will be, obviously, four yellow balls in bag 1 and x green balls and one yellow ball in bag 2. After the second transfer there will be either

a. Five yellow balls in bag 1 and x green balls in bag 2, or

b. Four yellow balls and one green ball in bag 1 and $(x - 1)$ green balls and one yellow ball in bag 2.

The probability that case a will exist is $1/(x + 1)$ and that case b will exist is $x/(x + 1)$.

After the third transfer there will be

a_1. Four yellow balls in bag 1 and x green balls and one yellow ball in bag 2 with a probability of $1/(x + 1)$—note that this distribution must occur if case a had occurred.

b_1. Four yellow balls in bag 1 and x green balls and one yellow ball in bag 2, with a probability of $\frac{1}{5}$ of $x/(x + 1)$.

b_2. Three yellow balls and one green ball in bag 1 and $(x - 1)$ green balls and two yellow balls in bag 2 with a probability of $\frac{4}{5}$ of $x/(x + 1)$.

If there is only one yellow ball in bag 2, the probability of drawing a green ball is $x/(x + 1)$. If there are two yellow balls in bag 2, the probability of drawing a green ball is $(x - 1)/(x + 1)$. Since these events are mutually exclusive, the probability of drawing a green ball from bag 2 is the sum of the probability of there being only one yellow ball in bag 2 times the probability of drawing a green ball when this event has occurred, plus the probability of there being two yellow balls in bag 2 times the probability of drawing a green ball when this event has occurred, or

$$\left[\frac{1}{(x + 1)} + \frac{x}{5(x + 1)}\right]\frac{x}{(x + 1)} + \left[\frac{4x}{5(x + 1)}\right]\frac{(x - 1)}{(x + 1)} = \frac{3}{5}$$

or
$$\frac{(5 + x)x}{5(x + 1)^2} + \frac{4x(x - 1)}{5(x + 1)^2} = \frac{3}{5}$$

or
$$5x + x^2 + 4x^2 - 4x = 3x^2 + 6x + 3$$

or
$$2x^2 - 5x - 3 = 0$$

or
$$x = \frac{5 \pm \sqrt{5^2 + 24}}{2 \times 2} = \frac{5 \pm 7}{4} = 3 \text{ or } -\frac{1}{2}.$$

There were originally three green balls in bag 2.

3. the odds are two to one

Let r be the number of red cards in portion A. Then $2r$ is the number of black cards in portion A. This leaves $(26 - r)$ red cards originally in portion B and $(25 - r)$ red cards after the transfer. Also the number of black cards in portion B is obviously $(26 - 2r)$. Since the ratio of red cards to black cards after the transfer is 2 to 1,

$$\frac{(25 - r)}{(26 - 2r)} = \frac{2}{1}$$

or
$$25 - r = 52 - 4r$$

Therefore
$$r = 9$$

There were originally twenty-seven cards in portion A (nine red and eighteen black) and twenty-five cards in portion B (seventeen red and eight black).

4. how many left?

You can go to a great deal of unnecessary work with this problem if you don't notice that the first and second transfer are meaningless except to mix you up.

The problem can be restated as follows: Given twenty-four marbles, of which eight are blue, eight red, and eight yellow, how many must you select at random so as to be certain of having at least three of each color?

The answer is obviously $8 + 8 + 3$, or 19. Five marbles were left in Bag 2.

5. tennis at lower dropshot

This is another of those cinch problems if you merely note that each match eliminated one player. Eighty-one matches were necessary to determine the club champion.

6. face cards

Assume a portion of an ordinary pack of playing cards contains n cards, of which c are face cards, then the probability that neither of two cards drawn at random will be a face card is

$$\frac{(n - c)}{n} \cdot \frac{(n - c - 1)}{(n - 1)}.$$

Under the conditions of the problem, this probability equals ½ or

$$\frac{(n - c)}{n} \cdot \frac{(n - c - 1)}{(n - 1)} = \frac{1}{2}.$$

This reduces to

$$n = \frac{1}{2}(4c + 1 \pm \sqrt{8c^2 + 1})$$

where n and c are integers and c is limited to the values 1 to 12 inclusive and n is greater than c. Trial will show that none of these values is possible except

$$c = 1 \qquad n = 4$$
$$c = 6 \qquad n = 21.$$

It is now obvious that the only way Green and Pink can hold portions of the pack with the necessary property, and at the same time have White's portion the smallest, is for both Green and Pink to hold twenty-one cards, of which six are face cards. This leaves only ten cards for White and, needless to say, none of these are face cards.

7. the king of hearts

At first glance you may wonder what effect, if any, all the hocus-pocus has on the basic fact that there are two kings of hearts among the 104 cards and that the chance that a particular card selected from the two packs is, under normal conditions, equal to $\frac{2}{104}$, or $\frac{1}{52}$.

But notice that one card in pack A is known to be the queen of hearts and thus not the king of hearts. It follows that the probability that the card transferred to pack B was the king of hearts is $\frac{1}{51}$ and not $\frac{1}{52}$.

It now follows that the probability that the card on the top of pack B is the king of hearts is the sum of two mutually exclusive probabilities—the probability that the transferred card was the king of hearts ($=\frac{1}{51}$) and that the top card is now one of the two kings of hearts in pack B, and the probability that the transferred card was not the king of hearts ($=\frac{50}{51}$) and that the top card is the only king of hearts in pack B. That is

$p = \frac{2}{53}$ times $\frac{1}{51}$ plus $\frac{1}{53}$ times $\frac{50}{51}$ (remembering that there are now 53 cards in pack B), or

$p = \frac{2}{2703} + \frac{50}{2703} = \frac{52}{2703}$ which is just a trifle greater than $\frac{1}{52}$.

8. birthdays

This probability is much larger than most people realize. The reason is that they confuse this with the probability that another individual will have the same birthday as one specially selected individual; this problem involves the probability that *any two* individuals will have the same birthday. The easiest way to determine this latter probability is to determine the probability that no two individuals have the same birthday and then subtract this value from 1.

The probability that A and B will have different birthdays is $^{364}/_{365}$. The probability that C will have a birthday different from A or B is $^{363}/_{365}$. Likewise the probability that D will have a birthday different from A or B or C is $^{362}/_{365}$, and so on. The probability that the sixth individual will have a birthday different from the preceding five is, of course, $^{360}/_{365}$.

The probability that all five of these events will occur simultaneously—that no two of the six individuals will have the same birthday, is the product of these five fractions, or

$$P(\text{none}) = \frac{364 \times 363 \times 362 \times 361 \times 360}{365 \times 365 \times 365 \times 365 \times 365} = 0.96$$

approximately. The probability that at least one pair of individuals out of six selected at random will have the same birthday is

$$P(\text{1 or more}) = 1 - 0.96 = 0.04 = 4 \text{ per cent,}$$

or one chance in twenty-five.

While one chance in twenty-five may not appear sensational, it is certainly not a rare event. It is interesting to note that the probability that two or more individuals in a room will have the same birthday increases with an increase in the number of individuals in the room much more rapidly than most people realize or suspect.

If there are twenty-three doctors gathered in a room the probability is more than 50 per cent that at least two have the same birthday. If there are forty in the group the probability is

$$P_{40} = 1 - \frac{364!}{365^{39} \times 325!} = 1 - 0.109 = 0.891.$$

In other words, the odds are more than eight to one that some two of the forty doctors in the group would have the same birthday!

9. the walk to 15th and m

Any particular route can be considered as being designated by
a combination of seven *N*s and seven *E*s. For example,

$$N\ N\ N\ E\ E\ E\ E\ N\ N\ E\ E\ N\ N$$

would indicate that he walked three blocks north, five blocks east,
two blocks north, two blocks east, and finally two blocks north
to his destination.

The number of different ways that he can travel then becomes
the number of possible permutations of fourteen items when there
are seven of one kind and seven of another. This is

$$\frac{14!}{7!\ 7!} = 3432$$

(see the introduction to this section).
There are 3432 possible routes.

10. the three prisoners

Ruling out the possibility of ties, there were originally six possible
orders of guilt, namely

$$A\text{-}B\text{-}C \quad A\text{-}C\text{-}B \quad B\text{-}A\text{-}C \quad B\text{-}C\text{-}A \quad C\text{-}A\text{-}B \quad \text{and} \quad C\text{-}B\text{-}A.$$

Of the six, the three where *B* is more guilty than *C* are eliminated by the guard's statement. This leaves the following three
possible cases, all of which are as far as *A* knows equally likely:

$$A\text{-}C\text{-}B \quad C\text{-}A\text{-}B \quad C\text{-}B\text{-}A.$$

Since *A* is the most guilty in only one of these three possibilities, his chance of being executed is one in three. (It is interesting
to note that the information given to *A* by the guard does not
affect the probability of *A*'s being judged the most guilty. This is
proper. As long as *A*'s case has not been considered by the judges,
his probability of being executed cannot be changed by anything
the judges decide about the relative guilt of the other two.)

11. the dictator

There are ten possible ways of selecting two individuals from a group of five. So far as A knows, any one selection is equally as likely to represent the two victims as any other except he knows that, since the gate was not closed, the possibility that A and B are the two victims has been eliminated. The nine remaining combinations are

A-C	B-C	C-D
A-D	B-D	C-E
A-E	B-E	D-E

Since three of the nine involve A, his chance of being shot is one in three and his chance of survival is two in three. (A of course runs the further risk that the dictator may have changed his mind and ordered all five to be shot!)

12. more tennis at lower dropshot

They agreed to play five matches.

If it is even money that a player will win m or more tosses of a coin, there will be exactly $2m - 1$ tosses.

This can be seen if you note that the chance of $2m - 1$ wins exactly equals that of $2m - 1$ losses (no wins at all); $2m - 2$ wins equals $2m - 2$ losses; and so on to the middle pair, which is m wins and $m - 1$ losses.

Here is the complete table for five matches:

Cut wins	0	1	2	3	4	5
Number of ways	1	5	10	10	5	1

Thus there are sixteen ways in which he can win three or more tosses and sixteen ways in which he can win fewer than three tosses.

13. cubes

It is obvious that the small cubes derived from each large cube will bear varying totals, which cannot be less than $1 + 2 + 3 = 6$ or more than $6 + 5 + 4 = 15$. But the totals of any particular set of small cubes will evidently depend upon the way in which its faces are numbered.

Let the six faces be *A*, *B*, *C*, *D*, *E*, and *F*. Then any one of the five—say *B*—may be opposite to *A*. The remaining four can be formed into three different pairs of opposites—*C*, *D* and *E*, *F*; *C*, *E* and *D*, *F*; *C*, *F* and *D*, *E*. Thus, for purposes of a problem of this type, the six faces of a cube can be numbered in fifteen different ways. We have, therefore, fifteen different sets of small cubes which may be obtained from any one large cube.

The following table shows, for each possible set of small cubes, the various *totals* that will be obtained:

Opposite faces					*Totals of*					
	6	7	8	9	10	11	12	13	14	15
65 43 21	.	.	.	1	3	3	1	.	.	.
65 42 31	.	.	1	1	2	2	1	1	.	.
65 41 32	.	.	1	2	1	1	2	1	.	.
64 53 21	.	.	1	1	2	2	1	1	.	.
64 52 31	.	1	.	2	1	1	2	.	1	.
64 51 32	.	1	1	1	1	1	1	1	1	.
63 54 21	.	.	1	2	1	1	2	1	.	.
63 52 41	1	.	.	3	.	.	3	.	.	1
63 51 42	1	.	1	1	1	1	1	1	.	1
62 54 31	.	1	1	1	1	1	1	1	1	.
62 53 41	1	.	1	1	1	1	1	1	.	1
62 51 43	1	1	.	.	2	2	.	.	1	1
61 54 32	.	1	2	1	.	.	1	2	1	.
61 53 42	1	.	2	.	1	1	.	2	.	1
61 52 43	1	1	.	1	1	1	1	.	1	1

Since it is a two-to-one chance that a cube drawn at random from the hat has a total of nine, the number of cubes in the hat must be of the form $3k$, where k is a positive integer. Since the number of cubes in the hat equals eight times the number of sons, *n*, minus 2, it follows that

$$8n - 2 = 3k \qquad \text{or} \qquad k = \frac{8n - 2}{3}.$$

Inspection readily shows that $n = 4$ is the smallest number of sons that there can be to have k as an integral number. In this case there would be thirty cubes in the hat. Of this thirty, ten would have a total of nine. Since the professor removed one small cube with a total of 9, there must have been eleven such cubes initially. With four sons this would require that three of them had used a marking plan which gave them each three small cubes with a total of 9. The three sons must have used the marking plan 63 52 41 (since this is the only plan which has three totals of 9). Also, note that this plan has no totals of 7 or of 14. The plan used by the fourth son must have had two small cubes with a total of 9 and at least one small cube with a total of 7. He must, therefore, have used marking plan 64 52 31 (since this is the only plan which meets these two requirements). As it also contains one small cube with a total of 14, there was only one small cube with a total of 14 in the hat.

14. king's town

If there are x players in a section, $x(x - 1)/2$ games are required for every contestant to play against every other player (see introduction to this section).

Let there be m players in Section A, then there will be $m(m - 1)/2$ games played in Section A and

$$\frac{(24 - m)(24 - m - 1)}{2}$$

games played in Section B. Since there are sixty-nine more games played in Section B than in Section A,

$$\frac{m(m - 1)}{2} + 69 = \frac{(24 - m)(24 - m - 1)}{2}$$

and $m^2 - m + 138 = m^2 - 47m + 552$

or $46m = 414$

or $m = 9.$

Assume Gambit won a games and tied b games, then

$$a + \tfrac{1}{2}b = 5\tfrac{1}{2} \quad \text{and} \quad a + b = m - 1 = 8$$

or
$$\tfrac{1}{2}b = 8 - 5\tfrac{1}{2} = 2\tfrac{1}{2}.$$
Hence
$$b = 5.$$

Gambit drew five games.

15. professor of ancient history

The simplest way to approach this problem is to consider all possibilities.

A	B	Answers	Probability
truthful	truthful	same	$\tfrac{3}{4} \cdot \tfrac{4}{5} = \tfrac{12}{20}$
truthful	lie	different	$\tfrac{3}{4} \cdot \tfrac{1}{5} = \tfrac{3}{20}$
lie	truthful	different	$\tfrac{1}{4} \cdot \tfrac{4}{5} = \tfrac{4}{20}$
lie	lie	same	$\tfrac{1}{4} \cdot \tfrac{1}{5} = \tfrac{1}{20}$

Given the fact that the two men made the same response, the probability that they are telling the truth is

$$\frac{\tfrac{12}{20}}{\tfrac{12}{20} + \tfrac{1}{20}} = \tfrac{12}{13}.$$

16. two spades

The chance that the first card drawn will be a spade is $\tfrac{7}{13}$ and the chance that the second card drawn will also be a spade is $\tfrac{9}{12}$. Since these are independent events the chance that both will be spades is $\tfrac{7}{13}$ times $\tfrac{9}{12}$, or $\tfrac{7}{26}$.

17. smithley's problem

Obviously, any one of the four natives could be lying or he could be telling the truth. There are, therefore, sixteen possible combinations of answers (truthful or false) that they could have made. These are listed here, along with the answer that Dah would have made in each instance:

Case	D	C	B	A	Statement Dah would have made under stated conditions
1	T	T	T	T	D asserts C denied B said A lied.
2	L	T	T	T	D asserts C denied B said A told the truth.
3	T	L	T	T	D asserts C denied B said A told the truth.
4	L	L	T	T	D asserts C denied B said A lied.
5	T	T	L	T	D asserts C denied B said A told the truth.
6	L	T	L	T	D asserts C denied B said A lied.
7	T	L	L	T	D asserts C denied B said A lied.
8	L	L	L	T	D asserts C denied B said A told the truth.
9	T	T	T	L	D asserts C denied B said A told the truth.
10	L	T	T	L	D asserts C denied B said A lied.
11	T	L	T	L	D asserts C denied B said A lied.
12	L	L	T	L	D asserts C denied B said A told the truth.
13	T	T	L	L	D asserts C denied B said A lied.
14	L	T	L	L	D asserts C denied B said A told the truth.
15	T	L	L	L	D asserts C denied B said A told the truth.
16	L	L	L	L	D asserts C denied B said A lied.

From the most unusual answer of Dah to Smithley's question, it is obvious that one of the following cases must be the true one: 1, 4, 6, 7, 10, 11, 13, or 16. The probabilities of these cases are as follows:

Case	Action of D	C	B	A	Probability
1	T	T	T	T	$\frac{1}{256}$
4	L	L	T	T	$\frac{9}{256}$
6	L	T	L	T	$\frac{9}{256}$
7	T	L	L	T	$\frac{9}{256}$
10	L	T	T	L	$\frac{9}{256}$
11	T	L	T	L	$\frac{9}{256}$
13	T	T	L	L	$\frac{9}{256}$
16	L	L	L	L	$\frac{81}{256}$

If Ah told the truth, the sum of the probabilities of cases 1, 4, 6, and 7 is $\frac{28}{256}$. If Ah told a lie the sum of the probabilities that Dah would have said what he did—cases 10, 11, 13, and 16—is $\frac{108}{256}$.

Since Dah did assert that Cah denied that Bah said that Ah lied, the probability that Ah told the truth is

$$\frac{^{28}\!/_{256}}{^{28}\!/_{256} + {}^{108}\!/_{256}} = \frac{28}{28 + 108} = {}^{28}\!/_{136} = {}^{7}\!/_{34}.$$

If you work out the probability that Bah, Cah, or Dah told the truth, the answer in the case of each one is also $^{7}\!/_{34}$. This leads to an apparent paradox. We know in advance that any one has a one in four chance of telling the truth. Yet here we find each one has only a seven in thirty-four chance.

The explanation lies in the fact that you have found by a process of elimination that either all told the truth, two lied and two told the truth, or all four lied. Hence the original condition of random truth telling one time in four no longer applies.

18. the odds against smith

Smith has a $^{1}\!/_{3}$ chance of drawing a bye. In this case he will have only one match to play and has, of course, a $^{1}\!/_{3}$ chance of winning this match and the championship. It follows that his chance of drawing a bye and then winning is $^{1}\!/_{3}$ times $^{1}\!/_{3}$, or $^{1}\!/_{9}$.

Smith also has a $^{2}\!/_{3}$ chance of not drawing a bye. In this case he will have to play, and win, two matches to win the championship. Since his chance of winning two successive games is $^{1}\!/_{3}$ times $^{1}\!/_{3}$, or $^{1}\!/_{9}$, his chance of not drawing a bye and then winning two games and the championship is $^{2}\!/_{3}$ of $^{1}\!/_{9}$, or $^{2}\!/_{27}$.

These two cases are mutually exclusive and Smith's total chance of winning the championship is therefore the sum of these two probabilities, or $^{1}\!/_{9} + {}^{2}\!/_{27} = {}^{5}\!/_{27}$.

While the original problem did not ask what were the chances of Brown's or Jones' winning, they can be calculated either in a similar manner or as follows:

The probability that either Brown or Jones will win is, obviously, one minus the probability that Smith will win, or $1 - {}^{5}\!/_{27} = {}^{22}\!/_{27}$. Since both Brown and Jones have an equal chance of winning, either one has a chance of winning equal to $^{1}\!/_{2}$ of $^{22}\!/_{27}$, or $^{11}\!/_{27}$.

19. professor knowsodds' reply

Suppose Professor A is the first one to take a wrong hat. Obviously, he had four ways of doing this since he had four equally desirable wrong hats to choose from. Let k be the number of ways the other four professors can pick up wrong hats, then the total number of ways for all five to get wrong hats will be $4k$.

But k is an integer and hence the total number of ways, $4k$, must be divisible by 4. Since none of the answers given, 50, 45, and 42, is divisible by 4, the professor knew instantly that they were all wrong.

20. opinionated is not satisfied

The easiest way to compute the number of ways in which none of the five professors will have his own hat is to subtract the number of ways that 1, 2, 3, 4, or 5 could have his correct hat from the total number of ways that five individuals could be given a hat. This latter number is, obviously, 5 factorial, or 120.

The computation of the other figures is as follows:

a. The number of ways in which all five get their own hat is, of course, one.

b. There is no way in which four can get their own hat and the fifth get someone else's hat.

c. There are ten ways in which the group can be subdivided into two groups, one subgroup consisting of three who have their own hats and one subgroup consisting of two who do not have their own hats. Since there is only one way in which any two individuals can receive the wrong hats, there must be ten ways in which five individuals can be given hats so that there are three, and only three, who receive their own hat.

d. There are ten ways in which the group can be subdivided into two groups, one subgroup consisting of two who have their own hats and one subgroup consisting of three who do not have their own hats. Since there are two distinct ways in which each of these groups of three could receive the

wrong hat, there must be twenty ways in which five individuals can be given hats so that there are two, and only
two, who receive their own hat.

e. There are five ways in which the group can be divided into
two groups, one subgroup consisting of one individual who
has his own hat and one subgroup consisting of four who
do not have their own hats. Since there are nine distinct
ways in which each of these groups of four could receive
the wrong hat, there must be forty-five ways in which five
individuals can be given hats so that there is one, and only
one, who receives his own hat.

There must be left $120 - 1 - 0 - 10 - 20 - 45 = 44$ ways
in which none of the five professors would receive his own hat.

21. a pair of dice

1. There are thirty-six combinations of two dice, since each
die can be any number from 1 to 6. Eleven of these combinations
will include at least one 4 (6–4; 5–4; 3–4; 2–4; 1–4; 4–6; 4–5;
4–3; 4–2; 4–1; and 4–4). When Joe announced that he did not
throw any 4s he has left only twenty-five possible rolls, one of
which is double 6s, so that the chance that Joe has rolled double
6s is $\frac{1}{25}$.

2. Similarly, there are eleven ways that Joe could have rolled
at least one 6. One of these is double 6s, so in this case the chance
that Joe rolled double 6s is $\frac{1}{11}$.

3. In this case we know that the first die is a 6. The chance
that the one which is still rolling will be a 6 is clearly $\frac{1}{6}$, so the
chance that Joe rolled double 6s, or at least will have rolled double 6s when the second die stops rolling, is $\frac{1}{6}$.

4. This problem is somewhat different from the other three in
that it involves inverse probability. The following assumptions
must be made about Joe:

a. That Joe has no intention of announcing anything more
than what one die showed.

b. That if the roll had been 6–5, 6–4, 6–3, 6–2, or 6–1, Joe would have been just as likely to state that he had rolled at least one 5, 4, 3, 2, or 1 as that he had rolled at least one 6.

On these assumptions we note that, while there were eleven possible rolls which would justify Joe's statement, the roll of double 6 must be given double weight since Joe could have made no other statement if he had rolled double 6s. We now have two favorable cases and ten unfavorable cases and the chance that Joe has rolled a double 6 is $\frac{2}{12}$ or $\frac{1}{6}$.

22. girls should live in brooklyn

It isn't a matter of the number of trains but rather of the time the trains stopped at that station. Obviously the schedule was such that the time between the departure of a Brooklyn-bound train and the next Van Cortlandt-bound train was about one-third the time between the departure of a Van Cortlandt-bound train and the next Brooklyn-bound train. As the time he arrived at the station was a random one, under these conditions it was to be expected that he would be three times as likely to find the next train Brooklyn-bound as heading in the other direction.

23. the three chess players

Obviously, the starting situation is unique. After the first game, each will be a contest between two men one of whom will have already won one game and if this player wins again he will be entitled to the prize. If he loses, the contest will continue with his opponent in the position of the man who has already won one game.

Let the chance of the one-game winner winning the series be X. The probability that he will lose and that his opponent will become the one-game winner is $\frac{1}{2}$. The chance of the opponent of the one-game winner winning the series is therefore $\frac{1}{2}X$. Similarly, the chance that both the one-game winner and his opponent will lose successive games and that the third man will become the one-game winner is $\frac{1}{4}$. The chance of the third man winning the series is thus $\frac{1}{4}X$. Since these three probabilities must add up to 1 it follows that

$$X + \tfrac{1}{2}X + \tfrac{1}{4}X = 1 \quad \text{or} \quad X = \tfrac{4}{7}$$

and the respective chances of the three players are $\frac{4}{7}$; $\frac{2}{7}$; and $\frac{1}{7}$.

The chance of the man who originally lost the toss is $\frac{2}{7}$ and the chances of the first two players, who face an equal probability of becoming the one-game winner or the third man, equals $\frac{1}{2}(\frac{4}{7}) + \frac{1}{2}(\frac{1}{7}) = \frac{5}{14}$.

The chances are then $\frac{5}{14}$, $\frac{5}{14}$, and $\frac{2}{7}$, the last being the chance of the player who sat out originally.

24. the five balls

You can go through a lot of mathematics to answer this problem but you don't have to. Just note that five is half of ten and that, if there are nine blue balls and one red ball in the box and you divide the ten balls into two portions and then select one of these portions, it is even money that you will select the group containing all blue balls and leave behind the one red ball in the other group of five balls.

25. how many balls?

Obviously, the two balls drawn by the professor must be both red, both blue, or one red and one blue. The probability of these various combinations, if there are r red balls and b blue balls in the bag is

Probability of two red balls =

$$P_{rr} = \frac{r(r-1)}{(r+b)(r+b-1)}$$

Probability of two blue balls =

$$P_{bb} = \frac{b(b-1)}{(r+b)(r+b-1)}$$

Probability of one red followed by one blue ball or the reverse =

$$P_{rb} = \frac{2br}{(r + b)(r + b - 1)}$$

Under the conditions the professor stated, the following relations must hold:

$$P_{rr} = 5P_{bb} \quad \text{and} \quad P_{rb} = 6P_{bb}.$$

Or, after substituting for the three probabilities their equivalent and multiplying both sides of the equations by $(r + b)(r + b - 1)$, we have

$$r(r - 1) = 5b(b - 1) \quad \text{and} \quad 2br = 6b(b - 1) \quad \text{or} \quad r = 3(b - 1).$$

Substituting this value of r in the first equation gives

$$3(b - 1)[3(b - 1) - 1] = 5b(b - 1)$$
or
$$3[3(b - 1) - 1] = 5b$$
or
$$9b - 9 - 3 = 5b$$
or
$$4b = 12$$
or
$$b = 3$$
and
$$r = 3(3 - 1) = 6$$

There are nine balls in the bag, three blue and six red.

It will be noted that in the above process both sides of the original equation were divided by a factor containing the unknown, namely, $(b - 1)$. Setting this factor equal to zero gives $b = 1$ as another solution to the original equation. Since the conditions of the problem require b to be at least two it follows that this value of b does not lead to a second answer to the Professor's question.

4 *where inference and reasoning reign*

Mathematics and logic are twins. For centuries mathematicians have attempted to prove that logic is a branch of mathematics—while the logicians have tried to prove that the reverse is the case. So far, neither side has been able to convince the other. They will probably arrive at a conclusion about the same time they solve the problem of which came first—the chicken or the egg.

You will find some thoroughly challenging mathematical problems in this section. They depend for their solution on inferential deductions and logical reasoning rather than on mere straightforward mathematical principles. They are grouped into one two-part section because while some (like the Smith–Jones–Robinson gem) can be solved by pure deductive processes, others depend upon pure reasoning. For example:

> Farmer Brown has three and seven-ninths haystacks in one corner of his field and two and two-thirds haystacks in the opposite corner of the same field. If he puts them all together, how many haystacks will he have?

Where does deduction stop and reasoning begin? Many problems will involve both processes, with a little mathematics often

84

thrown in for extra zest. (Incidentally, Farmer Brown will have one large haystack.)

By their very nature, the problems in Part 2 of this section, *Reasoning*, will have little in common and require many different approaches in their solution. On the other hand, true inference problems will have at least these four characteristics in common:

1. The answer is unique since the conditions given exclude all but one correct answer.

2. Trial-and-error methods may be required, but the nature of the trial and error is determined by mathematical or logical reasoning—not by guesswork.

3. There is always a "simplest" line of approach. In most of the problems that follow, this simplest line is as well concealed as possible. In fact, many of the more difficult problems will appear insoluble because there seems at first to be inadequate data.

4. In the problems which have a mathematical basis, the mathematics involved will be simple. Also, when used, this mathematical basis will be only part of the story and nonmathematical considerations, expressed or implied in the original statement of the problem, must ultimately be brought to bear upon it.

One method of solving inferential problems is to list all possibilities and then eliminate those that fail to meet the various conditions imposed until only one possibility is left. Obviously, this remaining possibility must meet all the given conditions or the problem has no solution. If the number of possibilities is so great that this method is impracticable, the first step should be to reduce the size of the field to a reasonable number of possibilities by logic and then to analyze the remainder.

There is hardly a better way to demonstrate an approach to problems of this type than to discuss what is probably the granddaddy of all inference problems, that of the brakeman, fireman, and engineer. This story, which went the rounds in the 1930s, has been expressed in various ways, but the essence is this:

The names, not necessarily respectively, of the brakeman, fireman, and engineer of a certain train are Smith, Jones, and Robinson. The three passengers on the train who happen to have the same names—Smith, Jones, and Robinson—will be referred to hereafter as Mr. Smith, Mr. Jones, and Mr. Robinson to distinguish them from the employees. We are informed that:

1. Mr. Robinson lives in Detroit.
2. The brakeman lives halfway between Chicago and Detroit.
3. Mr. Jones earns exactly $2000 a year (this was 1930).
4. Smith beat the fireman at billiards.
5. The brakeman's next-door neighbor, one of the passengers, earns exactly three times as much as the brakeman.
6. The passenger who lives in Chicago has the same name as the brakeman.

What is the name of the engineer?

Since there are only six possible combinations of the three railroad men and their occupations we shall start by listing them:

Case Number	1	2	3	4	5	6
Brakeman	s	s	j	j	r	r
Fireman	j	r	s	r	s	j
Engineer	r	j	r	s	j	s

We can immediately eliminate cases 3 and 5 from this table since our fourth condition was that Smith beat the fireman at billiards. Smith therefore cannot be the fireman, and cases 3 and 5 may be crossed out.

From condition 2 of the problem we see that the brakeman lives halfway between Chicago and Detroit. His next-door neighbor, one of the passengers, cannot live either in Chicago or in Detroit. He therefore cannot be Mr. Robinson (condition 1). Nor can he be Mr. Jones, because of conditions 3 and 5 ($2000 is not exactly divisible by 3). Hence the brakeman's next-door neighbor is Mr. Smith.

We see that Mr. Smith cannot live in Chicago or Detroit. Since Mr. Robinson lives in Detroit (condition 1), Mr. Jones must live in Chicago and the brakeman must be Jones (condition 6). Returning to the table, we now see that this fact will permit us to eliminate cases 1, 2, and 6 and leave us with case 4 as the only possible answer.

Therefore the engineer's name is Smith.

As a check on the solution we see that this result is consistent with all six of the original conditions. Incidentally, as a check on the quality of the problem we note that all six of the original conditions were used in arriving at the result.

part 1. inference

1. ho island

The Colorful Isles—Ho and Ha—are somewhere in the Pacific, thousands of miles from anywhere. Each is inhabited by one or more races of great interest to the problemist; while these races look alike, they differ in their reactions when strangers ask them questions in all respects but one—if the query is one to which they feel the questioner may know the correct answer (for example, "Is this a tree with green leaves?"), all races will simply ignore it and make no answer at all.

On Ho Island there are two races, known as the Good Guys and the Bad Guys. The Good Guys, asked a question, always tell the truth. Bad Guys always lie.

Smithley, visiting Ho Island, was introduced to three natives named Tom, Dick, and Harry and asked to determine the race to which each belonged. He asked each of the natives one question:

Q. "Tell me, Tom, is Dick a Good Guy?"
A. "Yes."
Q. "Dick, do Tom and Harry belong to the same race?"
A. "No."

Q. "Harry, what do you say about Dick? Is he a Good Guy?"
A. "Yes."

After hearing the answers Smithley turned to his friend and told him the race to which each islander belonged. What was his answer?

2. the island of ha

Smithley's next stop was Ha Island. Here dwelt not two races but three, traditionally known as Whites, Reds, and Pinks. Whites, when interrogated, always tell the truth. Reds invariably lie. Pinks alternately lie and tell the truth, but there is no means of knowing whether a Pink's answer to the first question he is asked will be veracious or not. The only certainty is that if his first answer is truthful, his second answer will be a lie and vice versa.

As Smithley and his White guide went along Low Street they met a group of three natives. As friend Smithley began to question one of them the guide told him in an aside that he was speaking to Mr. Pink, that the other two in the group were Mr. White and Mr. Red. He further stated that one of the three was a White, one a Red, and one a Pink, but that he did not know to which race they each belonged. "Mr. Pink," said Smithley, "are you a member of the Pink, the White, or the Red race?"

"I am a Pink, sir."
"And Mr. White?"
"He is a member of the White race."
"So Mr. Red is a member of the Red race?"
"Obviously."

Is Mr. Red the Red? If not, what is he?

3. bridge

Six men and their wives played bridge at three tables arranged in a row East to West, two men and two women at each table. No man played at the same table as his wife, but Mr. and Mrs.

Smith sat facing one another; so did Mr. and Mrs. Brown. In the play of the first hand of the evening the spades held by Mr. Robinson, Mr. Smith, Mr. Jones, and Mr. Green were

A K 9 8 3;   10 8 2;   Q 5 2;   Q J 10 9 7 6

respectively and Mrs. Thomas held six spades. One woman at each table was dummy and sat facing North. The play, apart from that of Mrs. Robinson, who almost revoked on the second round of spades, was of good class. Mrs. Robinson explained that her attention was distracted by her partner's having just dropped a card. Mrs. Green's partner made a grand slam in spades.

How were the players seated at each table?

4. the diners

Acorn, Beech, Cedar, Dogwood, and Elder and their respective wives recently dined together at a restaurant. The seats (at a circular table) were so arranged that the men and women alternated and each woman was three places distant from her husband.

Mrs. Cedar sat on Mr. Acorn's right. Mr. Elder sat two places to the left of Mr. Cedar, while Mrs. Elder sat two places to the right of Mrs. Beech.

Who sat on Mrs. Acorn's left?

5. "fur and feather" show

The Messrs. Parrot, Budgerigar, Canary, Hamster, and Rabbit were competitors in our local "Fur and Feather" show. Each took first prize in a class corresponding to the surname of one of the others and second prize in another of these classes.

Mr. Rabbit, for example, showed the winning hamster and took second prize with his canary, while the namesake of the exhibit for which Mr. Budgerigar took second prize showed the winning parrot.

Who owned the parrot that took second prize?

6. the dine-out club

"Yes," said the President of the Dine-out Club, "our club is, you may say, one happy family. There are only five members, hence all but one of us holds an important office. Mr. Dragon is my father-in-law. Mr. Phoenix is my brother-in-law. The Vice-President is going to marry my cousin. Mr. Minotaur's wife is the Secretary's sister. Please dine with me tonight. The Treasurer is coming; so are Minotaur and Phoenix."

I accepted his invitation and had a most pleasant evening. The next night three members of the club dined with me. One of them —Centaur—I met for the first time. The others were Griffin and Phoenix.

What office, if any, does each member of the club hold?

7. who was executed?

Five criminals appeared before Judge Jeffries for sentence. Their names, strange to say, were Libel, Fraud, Blackmail, Theft, and Murder—each the namesake of the crime with which one of the others was charged.

The namesake of the crime with which Blackmail was charged was himself charged with the crime of which the namesake was charged with murder; the namesake of the crime with which Murder was charged was himself charged with the crime of which the namesake was charged with fraud.

All the prisoners were found guilty and sentenced. Theft, for example, got seven years. Which was executed?

8. literary dinner party

Mr. and Mrs. Thomas gave a literary dinner party. They invited an essayist, a historian, a dramatist, a novelist, and the spouse of each of the four. The ten diners sat informally at a round table, men and women in alternate seats. No wife sat next to her husband.

The members of the party were well known to one another, except that Mrs. Alltalk had never before met the historian and Mrs. Cart had not met the essayist.

These facts were released concerning the seating arrangements:

1. The essayist's wife sat between the dramatist's husband and the historian.
2. The historian's wife sat on the left of the host.
3. Mr. Bluster sat between Mrs. Cart and the novelist.
4. Mr. Alltalk sat on the right of the hostess and on the left of Mrs. Bluster.
5. Mrs. Alltalk sat next to the novelist's husband.

Who sat on Mr. Daydreamer's right?

9. bridge at the greens'

At the Greens' family party, eight of those present sat down at two tables to play bridge. Those participating were Messrs. Black, Green, Pink, and White and their respective wives.

Mr. White's partner was his daughter. Mr. Pink was playing against his mother. Mr. Black's partner was his sister. Mrs. Green was playing against her mother. Mr. Pink and his partner had the same mother. Mr. Green's partner was his mother-in-law. No player's uncle was participating.

Which woman had been married at least twice and how were the two tables made up?

10. mr. smith is?

In our village we have a Mr. Carpenter, a Mr. Machinist, and a Mr. Smith. One is a carpenter, one a machinist, one a smith. None follows the vocation corresponding to his name.

Each is assisted in his work by the son of one of the others. As with the fathers, so with the sons; none follows the trade that corresponds to his name.

If Mr. Machinist is not a carpenter, what is the occupation of young Smith?

11. the squabble

"We had quite a squabble," writes my aunt, "about the children's tea party. Hank and Bob each wanted to sit between two girls. Gladys refused to sit next to Hank. Tom wanted to sit next to Amy, and Amy was agreeable—provided she sat at Tom's right. While all this argument was going on Charlie and Eileen, who just wanted to eat, sat down and started to demolish the pile of crumpets.

"Fortunately the two seats they selected were such that it was still possible for me to seat the other five at the table (which was circular) in accordance with their demands."

Where did Charlie and Eileen sit with relation to each other?

12. the new member cut in

"My name's Bill Smith," said the new member of the bridge club. "May I cut in?"

"You certainly may," replied one of the players, "provided you can figure out each of our names before we finish this rubber. My name is North and I am sitting North. The other players are named East, South, and West and our first names are Jim, Joe, Tom, and George. Joe is my partner at the table and Mr. West's partner in business. Tom is sitting to the left of Mr. South and George's partner is Mr. East."

Who sat where?

13. bella's gifts

Bella finds herself short of cash this Christmas. She has therefore sent to each of five friends the present she received from one of the others.

In no case have two gifts been interchanged, which means that if Kate's had gone to Julia, Julia's would not have gone to Kate. Laura's present has gone to the girl whose present has gone to Celia; and Ada's has gone to the girl whose gift (a pincushion) has gone to Laura.

Whose present has been sent on to Ada?

14. smithley back at ho island

The natives of Ho Island (remember that wonderful island in the Pacific where the two races, the Good Guys and the Bad Guys, are identical in appearance and differ only in their attitude toward the truth; Good Guys always tell the truth when asked a question—Bad Guys always lie?) are particularly fond of my friend Smithley and delight to challenge his ability to identify them. On his most recent visit, four of them came up to him and said, "To what race do we belong? Each of us will answer one reasonable question, provided you ask us all the same question."

Smithley thought a moment, smiled, then asked each of them in turn (they were standing in a circle with Smithley in the center), "Are you and your left-hand neighbor of the same race?"

The first native said "Yes," the second "Yes," the third "No," the fourth "No."

To what race did each belong?

15. bridge on ha island

You will recall that on mysterious Ha Island the Whites always tell the truth; the Reds always lie; and the Pinks tell the truth and lie alternately—a Pink's first answer may be truthful or false.

Four natives were playing bridge there and a visitor interrogated in turn South, West, North, and East. To each he put these questions:

(1) To which of the three races does your left-hand neighbor belong? (2) To which your partner? (3) To which your right-hand opponent?

Answers received were

from South: (1) Red, (2) Red, (3) Pink.
from West: (1) Pink, (2) White, (3) White.
from North: (1) Pink, (2) Pink, (3) Red.
from East: (1) White, (2) Red, (3) Red.

To which of the three races did each player in fact belong?

16. confusing?

"This is the card room," said Ferret, showing me around his club. "Those four chaps playing bridge are all people you'd like to meet. Henry and George are playing against Henry and Arthur. Thomas is the best of the four."

"And which is Thomas?"

"He's the chap on Thomas's left."

Not too lucid a statement, was it? I found afterward that Ferret uses his friends' surnames and given names indiscriminately. The four at the table were George Henry, Henry Thomas, Arthur George, and Thomas Arthur.

Who was Thomas Arthur's partner?

17. flowers

Seven residents in our village are avid exhibitors at the local flower show. Their names are Mr. Aster, Mr. Dahlia, Mr. Geranium, Mr. Phlox, Mr. Rose, Mr. Snapdragon, and Mr. Sweetpea.

Each is the namesake of a flower which one of the others grows.

Phlox is brother-in-law to the sweetpea-grower. Dahlia cannot bear roses. Snapdragon has never met the grower of geraniums. Rose and the dahlia-grower each married the other's sister. Sweetpea has never played cards in his life. Four of the seven, however, have organized a regular bridge four; each member of this four grows a flower which is another member's namesake. The geranium-grower and Dahlia are the exponents of rival bidding systems. Aster is an only child. Phlox detests bridge.

What flowers do the seven exhibitors grow?

18. king arthur

King Arthur and his Queen invited five Knights and their wives to dinner at the Round Table. At the dinner, no husband sat next to his wife, but each was separated from his wife by the same number of places. The Queen sat opposite Lady Camomile, while Sir Bogus sat three places to the Queen's left. Sir Asphodel was three places to Lady Bogus' right. Lady Eggs was two places from the Queen. Lady Asphodel sat opposite Lady Dachshund.

Assuming that no two men sat in adjacent chairs, draw a plan of the table showing how the twelve diners were seated.

19. outdoors

Mrs. Outdoors has three daughters: Anne, Betty, and Clare. One is yachting at Newport, a second is at Bar Harbor, and the third at Woodstock. One is playing tennis, one is yachting, and one is playing golf.

Anne is not at Newport, Clare is not at Bar Harbor, and the daughter who plays golf is not at Woodstock.

If the yachting enthusiast is not Clare, who is playing golf and where?

20. horsemen

Catkin, Jorkin, Lambkin, and Pipkin each named his favorite horse after one of the other three. In a recent trial, each rode one of the four horses. None, however, rode either his own horse or the horse named for him.

Lambkin rode Catkin's horse. The horse owned by Pipkin was ridden by the namesake of Lambkin's horse. The owner of Catkin rode Pipkin.

What is the name of the horse of which the rider of Lambkin is the owner?

21. muddle at mixwell

A curious situation exists among the tradespeople in the old-world village of Mixwell. Mr. Baker, Mr. Carter, Mr. Glover, and Mr. Smith follow (though not respectively) the callings of baker, carter, glover, and smith. In no case, in point of fact, is the vocation the same as the name.

Each of the four has an only son who is apprenticed to one of the others; his future calling, again, being different from his name.

A recent visit to Mixwell elicited some additional facts. Young Glover is engaged to the future smith's sister, while his father's sister is married to the smith. Mr. Baker is married to the glover's widowed mother. Mr. Carter has no daughter.

What are the vocations of each of the four tradespeople, and what are the prospective vocations of their sons?

22. five sons

"A lot of confetti and rice about," commented the visitor.

"And no wonder," said old Gloom, the sexton. "We've just had a big wedding. Five of 'em at one time. A real family affair."

"Meaning?" the visitor asked politely.

"Each of the men married a sister of one of the others. Each of the old 'uns has become a father-in-law twice over. For instance, Elsie Bridge married young Canasta."

"What! You mean that in each of the five families a brother and a sister were married at the same time?"

"Yes, and what's more, each of the five fellows enjoys playing a card game which is the same as one of the other's names——"

"No two of them enjoy the same game?"

"Correct! Also, none of them married a girl whose maiden name is the same as the game he enjoys. I can't tell you any more than that, except I do know:

1. that young Gin's fiancé was the bridge-player's sister,
2. that the euchre-player married Gladys Canasta, and
3. that the bridge-player married the euchre-player's sister."

Who married Miss Dit Gin?

23. the island of hi

Another of the Colorful Isles, the island of Hi, is inhabited solely by Pinks. There are, however, three separate strains or subspecies. All, when asked a series of questions, tell the truth and lie alternately. But whereas Corals answer the first question truthfully, Roses begin by lying. The third strain, the Salmons, are like the Pinks elsewhere—their first answer may be either true or false.

Three natives who were being questioned as they sat at a circular table included one of each subgroup. Smithley, moving around in a clockwise direction, put three questions to each. These were: (1) What is your name? (2) What is the name of your left-hand neighbor? (3) To what race does he belong?

The first native replied: (1) "Pedro," (2) "Rodrigo," (3) "He's a Rose." The second: (1) "Rodrigo," (2) "Quixote," (3) "He's a Coral." The third: (1) "Pedro,' (2) "Rodrigo," (3) "He's a Rose." Did the Salmon answer his first question truthfully?

24. david?

Albert and Barbara and Charles and David are the cause of much confusion in our village. Each of them owns a cat named after one of the other three and a dog named after another of them. No two cats, and no two dogs, have the same name.

For example, David's dog and Charles' cat are both namesakes of the owner of the cat Charles. The namesake of Barbara's cat is the owner of the cat whose namesake owns the dog Albert.

Who owns the dog David?

25. the island of hu

Our wandering Smithley next proceeded to the Island of Hu. Here there are three races. The Whites (like the Whites elsewhere in these Colorful Isles) always give truthful answers to questions. The Pinks, like the Pinks of Ha Island, answer truthfully and mendaciously to alternate questions and may or may not step off on the honest foot.

The third race, the Greens, are a peculiar and proud race. Asked three or four questions, a Green will tell the truth exactly once. Moreover, he will answer only if you tell him in advance whether you are going to ask three or four questions. The reason for this requirement is that he does not want ever to be mistaken

for either a White or a Pink. Therefore, if asked only three questions, he always lies in answering the second one since if his three
answers are false–true–false he might be mistaken for a Pink—
something which would make him blush in shame (and thereby
completely lose face).

Mr. Smithley was introduced to three natives who bore the
(for him) nostalgic appellations of Smith, Jones, and Robinson.
He was invited to put three questions to each of them.

After profound thought he decided to ask each the same three
questions, which in effect were as follows:

1. Do Smith and Jones belong to the same race?
2. Do Jones and Robinson belong to the same race?
3. Do Robinson and Smith belong to the same race?

The answers he received were:

from Jones: (1) Yes, (2) No, (3) No.
from Smith: (1) No, (2) Yes, (3) No.
from Robinson: (1) Yes, (2) No, (3) Yes.

To which race do each of the three natives belong?

26. artichoke, bergamot, and celery

The three young race horses, Artichoke, Bergamot, and Celery,
are owned by three breeders after whom they are named. No
owner has the same name as his horse and each horse has been
trained by the breeder who is neither his owner nor his namesake.

In a recent trial gallop each of the breeders rode one of the
three horses. Captain Artichoke rode the horse trained by the
namesake of his own horse. Mr. Bergamot rode the horse which
he himself had trained.

Which horse was ridden by Charlie Celery?

27. the bank manager is?

Six friends, whose names were White, Red, Yellow, Green, Blue,
and Black, and whose occupations (though not necessarily respectively) were those of doctor, architect, lawyer, bank manager,

stockbroker, and engineer, dined together recently in the city. They were seated at equal intervals around a circular table. White sat on the doctor's left, Yellow on the architect's right, and Black opposite the engineer. During dinner, conversation turned to the subject of motoring; each of the diners, it transpired, possessed a car; but the six cars were all of different makes. Red had a Lincoln, the stockbroker a Chevrolet, the man opposite Yellow a Plymouth, the man on Green's left a Buick, the man on Blue's right a Chrysler, and the man opposite the lawyer a Ford. The owner of the Chrysler was on Red's left, the owner of the Chevrolet on Green's right, and the owner of the Buick opposite Blue.

What was the name of the bank manager?

28. white and green

Mr. and Mrs. White have two sons and two daughters; their names (not necessarily respectively) are Alison, Leslie, Hilary, and Sidney. Mr. and Mrs. Green have also two sons and two daughters, who, as it happens, have the same four names.

Moreover, each of the four Whites is engaged to one of the four Greens; no girl, however, has the same name as her fiancé. Miss Hilary Green is betrothed to the namesake of Sidney White's girl. Leslie White is a year older than her prospective husband.

Who is marrying Leslie Green?

29. strange!

Stranger even than the unusual case considered in problem number 7 in this section, "Who Was Executed," is the following unprecedented case which came before the Court of Criminal Appeal:

The five prisoners whose cases came up for review bear the somewhat unusual names of Arson, Burglary, Counterfeiting, Dognaping, and Embezzlement. Each of the five crimes corresponding to their names has been committed—or is alleged to have been committed—by one of the prisoners who is not the namesake of his crime. But, owing to some confusion in the courts below, the crime of which each has been convicted is in no case either that which he is alleged to have committed and with which he was originally charged, or that of which he is the namesake. For example, Burglary was charged with the crime for which Dognaping has been sentenced and Dognaping with the crime for which Counterfeiting has been sentenced. The namesake of this crime was charged with the crime of which the namesake was charged with counterfeiting. All the prisoners are interested in this appeal, but especially Arson, who had been sentenced to thirty years for embezzlement.

Lord Justice Bloodhound, famous for his quick-wittedness, quickly deduced which of the five prisoners had been sentenced for dognaping. Who was this?

solutions

1. ho island

1. Since Tom and Harry answered the same question identically they must be of the same race.

2. Dick was therefore lying when he said they were not the same race and must be a Bad Guy.

3. Since both Tom and Harry said Dick was a Good Guy, they also answered untruthfully.

4. All three are Bad Guys.

An interesting question might be asked about this problem. "Would Smithley have been able to determine the race of the natives, regardless of what they actually were, from the questions he asked?" The following table gives the answers that would be given by the natives for the eight possible race combinations:

RACE COMBINATION

Name	1	2	3	4	5	6	7	8
Tom	good	good	good	good	bad	bad	bad	bad
Dick	good	good	bad	bad	good	good	bad	bad
Harry	good	bad	good	bad	good	bad	good	bad

Question				Answers				
Dick Good?	yes	yes	no	no	no	no	yes	yes
T & H same race?	yes	no	no	yes	no	yes	yes	no
Dick Good?	yes	no	no	yes	yes	no	no	yes

Smithley's questions were entirely adequate to identify the race to which each of the three belonged, regardless of the true situation. Each possible race combination has its own distinctive pattern of answers.

2. the island of ha

1. Mr. Pink is not a member of the White race, for if he were, he would say so.

2. Mr. Pink is not a member of the Pink race. For, if he is, his first answer is truthful and hence his third answer must also be truthful. If this is the case it is evident that his second answer is also truthful; which is impossible.

3. So Mr. Pink is a member of the Red race. All his answers are untruthful. Mr. White is the Pink and Mr. Red the White.

4. Mr. Pink is a Red.
   Mr. White is a Pink.
   Mr. Red is a White.

3. bridge

1. a. Mr. Robinson held   A K 9 8 3   of spades.
   b. Mr. Smith held   10 8 2   of spades.
   c. Mr. Jones held   Q 5 2   of spades.
   d. Mr. Green held   Q J 10 9 7 6   of spades.
   e. Mrs. Thomas held   six spades.
   f. Mrs. Robinson held   at least two spades.

2. Mr. S cannot sit at the same table with Mr. R (both hold the 8 of spades), Mr. J (both hold the 2 of spades), or Mr. G (both hold the 10 of spades). Similarly, Mr. G cannot sit with Mr. R (both hold the 9 of spades), or with Mr. J (both hold the Q of spades). It follows that Mr. R and Mr. J are at one table and Mr. S and Mr. G at the second and third tables.

3. Mr. and Mrs. S and Mr. and Mrs. B are sitting East and West, since each wife faced her husband.

4. At the table at which Mr. R and Mr. J sat, either Mrs. G or Mrs. T had to be sitting South. However, since Mr. R holds five spades and Mr. J three, it is not possible for Mrs. T (who holds six spades) to be at the same table with them. Mrs. G is sitting South at the JR table.

5. Mrs. G's partner, who made a grand slam in spades, has to be Mr. R (a grand slam in spades against Mr. R's holding is impossible). So Mr. R is sitting North. This requires Mr. J to be sitting West (if he were sitting East behind Mr. R, his Q of spades would take a trick and make a grand slam in spades impossible). Note that the play is described as being of good class and that it is only possible to finesse against the Q 5 2 if they are in front of the A K 9 8 3.

6. Mrs. R sat either East or West, since she almost revoked on defense and North played all hands. However, since Mr. J is sitting West, Mrs. R must be sitting East (see 3 above).

7. This leaves as the four North-South players at the other two tables Mr. and Mrs. T, Mr. G and Mrs. J. Since no husband plays at the same table with his wife, Mr. G sat North and played with Mrs. T and Mr. T sat North and played with Mrs. J.

8. Since Mr. G and Mrs. T hold twelve spades between them, Mrs. R (who holds at least two spades) cannot be sitting at their table and must be sitting East against Mr. T and Mrs. J (she cannot sit at the same table with Mr. R).

9. The above information is sufficient to definitely locate eight players at the tables:

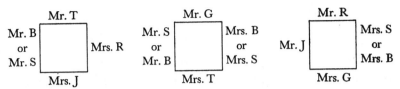

The table arrangement shown above may not be obvious at first glance, but a little reflection will show that it is the only arrangement which will fit the above facts and still permit Mr. and Mrs. S and Mr. and Mrs. B to sit East and West (3 above) and still have each husband and wife facing each other.

10. Obviously, Mr. S (who holds three spades) cannot be sitting at the same table with Mr. G and Mrs. T (who hold twelve spades) or with Mr. R and Mr. J (this would put three men at the same table). He must then be sitting West at the table with Mr. T. This fact gives the following as the exact seating arrangement:

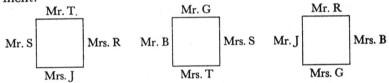

4. the diners

1. Trial will quickly show that either each woman will be three places to the right of her husband or each woman will be three places to the left of her husband. With five couples there is no possible arrangement in which some of the women are three places to the right, and others three places to the left, of their husbands. If we represent men by capital letters and women by lower-case letters we must have one of the following two arrangements (the names are listed in a counterclockwise order, that is, in the first arrangement Mrs. V is on Mr. R's right, Mr. S is on Mrs. V's right, ———, Mr. R is on Mrs. U's right):

a. When the wife is three places to the right of her husband—

$$R \quad v \quad S \quad r \quad T \quad s \quad U \quad t \quad V \quad u \quad R$$

(R is listed twice to indicate the circular seating).

b. When the wife is three places to the left of her husband—

$$R \quad t \quad S \quad u \quad T \quad v \quad U \quad r \quad V \quad s \quad R.$$

2. Note that in both seating arrangements the order of the men and of the women is identical. Therefore, since Mrs. Elder sat two places to the right of Mrs. Beech, Mr. Elder must have sat two places to the right of Mr. Beech.

3. Taking into account the additional fact that Mr. Elder sat two places to the left of Mr. Cedar, our two seating arrangements become

a. $R \quad v \quad B \quad r \quad E \quad b \quad C \quad e \quad V \quad c \quad R$
b. $R \quad e \quad B \quad c \quad E \quad v \quad C \quad r \quad V \quad b \quad R.$

4. Of these two arrangements, in the first one Mrs. Cedar is on the right of Mr. V and in the second one on the right of Mr. Beech. Under the conditions of the problem the latter is impossible. Therefore, the first is the only possible seating arrangement and V is Acorn, leaving R to be Dogwood and giving

$$D \quad a \quad B \quad d \quad E \quad b \quad C \quad e \quad A \quad c \quad D.$$

Mr. Dogwood sat on Mrs. Acorn's left.

5. "fur and feather" show

1. The exhibitor who took first prize with the winning parrot cannot be Mr. Parrot (same name), Mr. Canary (Mr. Rabbit took second prize with his canary and hence Mr. Budgerigar could not have also taken second prize with a canary), Mr. Budgerigar (as he would then have won second prize with a budgerigar), or Mr. Rabbit (who took first prize with his winning hamster). Mr. Hamster must have won first prize with his winning parrot.

2. The owner of the parrot which took second prize cannot be Mr. Rabbit (who took a second prize with his canary), Mr. Budgerigar (who took a second prize with his hamster), Mr. Hamster (he took a first prize with his winning parrot), or Mr. Parrot (same name).

3. Mr. Canary must have owned the parrot that took second prize.

6. the dine-out club

1. The President is clearly not *D*, *P*, or *M*. Nor can he be *C* (who was not present the first night). Therefore he must be *G*.
2. The Treasurer is not *M* or *P*. Neither can he be *G* (the President) nor *C* (who was not present the first night). Therefore the Treasurer is *D*.
3. *M* is not the Treasurer, President (these offices are already filled), Secretary (he is married to the Secretary's sister), or Vice-President (since the Vice-President is not married). Hence he holds no office.
4. Either (*a*) the President's wife is *P*'s sister, or (*b*) *P*'s wife is the President's sister. But (*a*) cannot be true (the President's father-in-law is *D*). Thus (*b*) is true and therefore *P* is married and cannot be the Vice-President; he must be the Secretary and *C* the Vice-President.
6. To summarize:

> Griffin—President
> Centaur—Vice-President
> Dragon—Treasurer
> Phoenix—Secretary
> Minotaur—no office.

7. who was executed?

1. Blackmail → *x*; Mr. *X* → *y*; Mr. *Y* → murder.
2. Murder → *z*; Mr. *Z* → *w*; Mr. *W* → fraud.
3. Neither Murder (same name), Blackmail (*1* above), nor Theft (who got seven years) could have been charged with murder.
4. Assume that Fraud is charged with murder. Then *y* = fraud, and we have: Blackmail → *x*; Mr. *X* → fraud; Fraud → murder.

5. Combining *4* with *2* above, we see that $x = w$ and $b = z$. This gives us: Murder → blackmail; Blackmail → x; Mr. X → fraud; Fraud → murder. This is a closed cycle involving only four individuals and leaves the fifth to be charged with the crime of which he is the namesake. Since this cannot be, it follows that Fraud cannot be the one charged with murder and that Libel must have been hanged for murder.

NOTE: If desired the problem can be extended to ask "What crime did each criminal commit?" The rest of the answer would be

6. Requirement *1* can be rewritten as follows:

Blackmail → x; Mr. X → libel; Libel → murder.

7. The only way in which this sequence can be combined with the requirement in *2* above is for $x =$ fraud. This in turn will make $w =$ blackmail and leave $z =$ theft, thus giving as the final answer:

Murder → theft; Theft → blackmail; Blackmail → fraud; Fraud → libel; Libel → murder.

8. literary dinner party

1. Mrs. Daydreamer must be the novelist (clue 3 eliminates Cart and Bluster; clue 5 eliminates Alltalk).
2. Mrs. Alltalk must be the dramatist (she has never met the historian; Mrs. Daydreamer is the novelist; the essayist's wife is sitting between the dramatist's husband and the historian according to clue 1, and from clue 5 we know that Mrs. Alltalk sat next to the novelist's husband).
3. Mr. Cart is, therefore, the historian (Mrs. Cart has never met the essayist) and Mr. Bluster the essayist.
4. Given that Mrs. Alltalk is the dramatist, Mr. Bluster the essayist, Mr. Cart the historian, and Mrs. Daydreamer the novelist, it is easy enough to establish the seating arrangement as follows:

Hostess	Alltalk (on the hostess' right)		Mrs. Bluster
Cart	Mrs. Daydreamer	Bluster	Mrs. Cart
Host	Mrs. Alltalk	Daydreamer	Hostess (listed

twice to indicate the circular arrangement)

The hostess, Mrs. Thomas, therefore sat on Mr. Daydreamer's right.

9. bridge at the greens'

1. Mr. *W* is playing with his daughter; Mr. *B* with his sister; and Mr. *G* with his mother-in-law. Each set of partners accordingly consists of a man and a woman.

2. Mr. *G* is playing with his mother-in-law, and Mrs. *G* is playing against her mother. Therefore they are playing at the same table against each other.

3. Mr. *P* is not playing with Mrs. *P* (Mr. *P* and his partner have the same mother) or against Mrs. *P* (Mr. *P* is playing against his mother). It follows that Mr. and Mrs. *P* are playing at different tables.

4. If Mr. *P* and Mr. *B* are at the same table, Mr. *B*'s partner (his sister) is Mr. *P*'s mother and Mr. *B* his uncle. But no player's uncle was participating. Therefore, Mr. *B* and Mrs. *P* are at one table and Mr. *P* at the other.

5. If Mr. *B* and Mrs. *P* are at the same table with Mr. and Mrs. *G*, Mr. and Mrs. *W*, Mr. *P* and Mrs. *B* must be at the other table. In this case Mr. *W* must be playing with Mrs. *B*, his daughter, against Mr. *P* and Mrs. *W*, or:

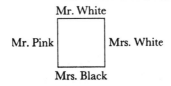

Mr. White

Mr. Pink | Mrs. White

Mrs. Black

Since Mr. *P* and his partner have the same mother and Mr. *P* is playing against his mother, Mrs. *B* is the mother of both Mr. *P* and Mrs. *W*. This in turn means that Mrs. *W* is Mr. *W*'s grand-daughter. But Mr. *W* cannot be married to his own grand-daughter, hence this table arrangement will not fit the conditions of the problem and Mr. *B* and Mrs. *P* must be at one table, say table number one, and Mr. and Mrs. *G* and Mr. *P* at table number two. This gives us (since, obviously, Mr. *W* cannot sit at table two with Mr. *P* and Mr. *G*):

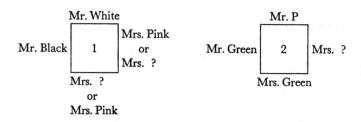

6. If Mrs. *W* is at table one she will be playing with Mr. *B* (Mr. *W* is playing with their daughter). Mrs. *W* is, therefore, Mr. *B*'s sister. This would make Mr. *B* an uncle, contrary to the condition that no one playing is an uncle, hence Mrs. *W* must be sitting at table two with Mr. *G*.

7. Now all we have to do is to locate Mrs. *P* and Mrs. *B* at table one. Here Mr. *B* is playing with his sister, so he can't be playing with Mrs. *B*, and must be playing with Mrs. *P*.

8. This gives us the complete seating:

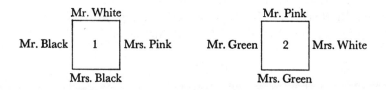

9. Since Mrs. White is the mother of Mr. Pink, Mrs. Green (whose maiden name is not given), and Mrs. Black (whose maiden name was Miss White), she (Mrs. White) must have been married at least twice.

10. mr. smith is?

Mr. Machinist cannot be a machinist or a carpenter. He must therefore be a smith, Mr. Carpenter a machinist, and Mr. Smith a carpenter.

Young Smith, who cannot assist his father or the man whose trade is that of smith, must assist Mr. Carpenter and be a machinist.

11. the squabble

Start with Tom with Amy on his right. Since there are only three girls and both Bob and Hank wish to sit between two of them, one of these two must come next. Obviously it has to be Hank as otherwise he would be next to Gladys. After Hank there must be Eileen, followed by Bob and Gladys. Naturally the remaining empty seat between Gladys and Tom is occupied by Charlie, the complete order being

Tom → Amy → Hank → Eileen → Bob → Gladys → Charlie → Tom (Tom's name is listed twice to indicate the circular arrangement). It is evident that Charlie sat down three places to Eileen's right.

12. the new member cut in

Since Joe was Mr. North's partner at the table and Mr. West's partner in business he had to be either Mr. East or Mr. South.

If Joe were Mr. East, then Mr. North's first name would be George. North and East would be partners and Tom would have to be at the left of one of them. But Tom is sitting to the left of Mr. South, so this is impossible.

So Joe must be Joe South, and sitting South.

Tom is sitting to the left of Joe South; his last name must be East since George's partner is Mr. East. That makes George's last name West; the whole arrangement finds Jim North sitting North, George West sitting East, Joe South sitting South, and Tom East sitting West.

13. bella's gifts

1. There are five gifts to be exchanged.

2. Four gifts cannot be circulated among the four girls who gave them originally; this would leave the fifth girl to receive the gift she herself gave.

3. Likewise, three gifts cannot be circulated among the three girls who gave them originally because this would leave two girls who would have either to interchange gifts or to receive back the gifts which they had originally given—both of which are prohibited by the conditions of the problem.

4. In other words, the exchange of gifts must have involved all five girls and have been of the following type (*A* appears twice to indicate that it is a circular arrangement):

$$A \rightarrow B \rightarrow C \rightarrow D \rightarrow E \rightarrow A.$$

5. If we let *A* be Ada and insert our known data, we have:

$$\text{Ada} \rightarrow x \rightarrow \text{Laura} \rightarrow y \rightarrow \text{Celia} \rightarrow \text{Ada}.$$

6. Thus, while we cannot give the complete distribution we can definitely state that Celia's gift was sent on to Ada.

14. smithley back at ho island

Only four possible situations exist so far as any native being questioned is concerned:

1. Good Guy with a Bad Guy on his left.
2. Good Guy with a Good Guy on his left.
3. Bad Guy with a Bad Guy on his left.
4. Bad Guy with a Good Guy on his left.

Adding the reply the native would make in each of these cases gives

1. Good Guy with a Bad Guy on his left would say "No."
2. Good Guy with a Good Guy on his left would say "Yes."
3. Bad Guy with a Bad Guy on his left would say "No."
4. Bad Guy with a Good Guy on his left would say "Yes."

Notice that, regardless of the race of the native being questioned, the native on his left is a Bad Guy if he answers "No" and is a Good Guy if he answers "Yes."

The first native's answer therefore told Smithley that the native on his left was a Good Guy. The second native's reply indicated that the native on his left was also a Good Guy. The third and fourth natives' replies indicated that those on their left were both Bad Guys.

15. bridge on ha island

Since the Whites always tell the truth, the Reds always lie, and the Pinks alternately tell the truth and lie, if anyone's first statement is false so is his third, and vice versa. Set out here, therefore, are the several statements made, with the first and third statement being placed in italics for ready reference:

	Statements about			
	WEST	NORTH	EAST	SOUTH
by South	*Red*	Red	*Pink*	
by West		*Pink*	White	*White*
by North	*Red*		*Pink*	Pink
by East	Red	*Red*		*White*

West's third statement is that South is a White. But if South is a White all of his statements must be true and West is a Red. This leads to a contradiction since West's third statement cannot then be a true one. South is therefore not a White and, since West's first statement must also be a false one, North is not a Pink.

Since South was not a White, East's first answer (and hence his third) was false. Therefore, North is not a Red.

Since North is not a Pink or a Red he must be a White and all his answers true. Therefore

> West is a Red
> East is a Pink (lying the first and third times)
> South is a Pink (lying the second time)
> North is a White

16. confusing

The two Henrys and the two Thomases were on opposite sides. Therefore, Henry Thomas was opposed by Thomas Arthur and George Henry, and thus Thomas Arthur's partner was George Henry.

17. flowers

1. In view of the fact that the geranium-grower, who plays bridge, has never met Mr. Snapdragon it follows that Snapdragon is not one of the bridge foursome. Messrs. Phlox and Sweetpea obviously are the other two non-bridge players.

2. Among the bridge players—each of whom grows a flower which is another player's namesake—Mr. Dahlia detests roses and opposes the geranium-grower at bridge. He must therefore grow asters. Mr. Rose is not the dahlia-grower (they married each other's sister) or the aster-grower (grown by Dahlia). It follows that he grows geraniums. Mr. Aster has no sister and thus cannot be the dahlia-grower. This leaves Aster as the rose-grower and Geranium as the dahlia-grower.

3. Among the non-bridge players, since Mr. Phlox cannot be the sweetpea-grower (his brother-in-law) or the phlox-grower he must be the snapdragon-grower. This leaves Sweetpea as the phlox-grower and Snapdragon as the sweetpea-grower.

To summarize:

> Dahlia grows asters.
> Rose grows geraniums.
> Aster grows roses.
> Geranium grows dahlias.
> Phlox grows snapdragons.
> Snapdragon grows sweetpeas.
> Sweetpea grows phlox.

18. king arthur

1. Using capital letters to represent men, lower-case letters to represent women, and numerals to represent unknown positions, we may represent the information given in the first clue as fol-

lows (the positions indicated in counterclockwise order; that is, Sir 2 is on Lady Camomile's right, Lady 3 on Sir 2's right, Sir Bogus on Lady 3's right, — — —, Lady Camomile on Sir 12's right):

$$c \to 2 \to 3 \to B \to 5 \to 6 \to q \to 8 \to 9 \to 10 \to 11 \to 12 \to c$$

(c is shown at both ends of the listing to indicate the circular seating).

2. Since the Queen is opposite Lady Camomile and Lady Asphodel is opposite Lady Dachshund, Lady Bogus must be opposite Lady Eggs. Lady Eggs, who sits two places from the Queen, must be in position number 5 or number 9. If Lady Eggs is in position number 9 Lady Bogus will be in position number 3— next to Sir Bogus. Since this is ruled out under the conditions of the problem, it follows that Lady Eggs is in position number 5, Lady Bogus in position number 11, and Sir Asphodel, who is three places to Lady Bogus' right, is in position number 2. This brings us to:

$$c \to A \to 3 \to B \to e \to 6 \to q \to 8 \to 9 \to 10 \to b \to 12 \to c$$

3. It remains a simple matter to complete the seating arrangement as follows (since each husband is separated from his wife by the same number of places and Sir Bogus is five places to Lady Bogus' right):

$$c \to A \to d \to B \to e \to C \to q \to D \to a \to E \to b \to K \to c$$

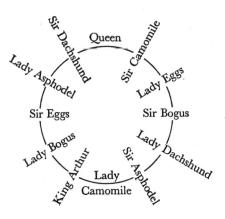

19. outdoors

The girl who is at Newport is yachting. Anne is not at Newport and, since Clare is not a yachting enthusiast, it follows that Betty must be yachting at Newport.

As Clare is not at Bar Harbor, Anne must be.

So Clare is at Woodstock and thus is not playing golf.

Therefore, Anne is playing golf at Bar Harbor.

20. horsemen

The data can be tabulated as follows.

HORSE	OWNER	RIDER
	C	L
m	L	
	P	m
C	n	
P		n

Since we have no information about Jorkin, the letter J does not appear. The first row shows that Lambkin rode Catkin's horse. The second and third show that the rider of Pipkin's horse had the same name as Lambkin's horse; the letter m may be either Jorkin or Catkin.

Similarly, n may be the name of any owner except Catkin or Pipkin.

Trying out the two possibilities for m, we see that m cannot be J because in that case n would have to be L. L owns horse m and L would also own horse C—which is contrary to the conditions of the problem.

So m is C; n becomes L, and the complete table is

HORSE	OWNER	RIDER
P	C	L
C	L	J
J	P	C
L	J	P

The rider of Lambkin is Pipkin, who owns the horse named Jorkin.

21. muddle at mixwell

1. The future smith is neither young Smith nor young Carter (for old Carter has no daughter), nor young Glover (who is engaged to the future smith's sister).

The future smith is therefore young Baker.

2. Old Baker is not the glover (whose widowed mother he married), or the smith (young Baker is the future smith), or, of course, is he the baker.

Old Baker is the carter.

3. Old Glover is not the carter, or the glover, or the smith—whose wife is old Glover's sister.

Old Glover is the baker.

The remaining facts can now be deduced at sight, the final table being:

VOCATIONS

Name	Old	Young
Baker	carter	smith
Carter	smith	glover
Glover	baker	carter
Smith	glover	baker

22. five sons

1. Let B, C, E, G, and P represent the five young men and b, c, e, g, and p the five games. Placing the given data in tabular form gives

Man	Married	Enjoys
B		
C	Miss B	
E		
G	Miss X	
P		
X	Miss Y	b
Y	Miss C	e

2. *Y* is not *C* (he married Miss *C*), *E* (the euchre-player), or *B* (Miss *Y* married the bridge player). Therefore, *Y* is *G* or *P*.

3. *X* is not *Y* (he married Miss *Y*), *G* (Miss *X* is married to *G*), *B* (Miss *B* is married to *C* and cannot also be married to *G*), *C* (if *X* is *C* then *Y* is *B*, and both *B* and *G* would be married to Miss *C*), or *P* (if *X* is *P* then, from 2 above, *Y* is *G* and both Miss *P* and Miss *C* will be married to *G*). It follows that *X* is *E*.

4. Returning to *Y*, if *Y* is *G* and *X* is *E* (3 above), *G* will be married to both Miss *C* and Miss *E*. Therefore *Y* must be *P*.

5. Substituting these values in the table gives

Man	Married	Enjoys
B		
C	Miss B	
E	Miss P	b
G	Miss E	
P	Miss C	e

6. *B* married Miss *G*.

7. The gin player cannot be *B* (*B* married Miss *G*), *G* (same name), *E* (*E* plays bridge), or *P* (*P* plays euchre). He is, therefore, *C*.

8. This leaves *B* and *G* as the canasta and poker players, and we cannot tell which one is which; but we can say that Mr. Bridge married Miss Dit Gin.

23. the island of hi

Coral's first and third answers are both truthful. And one of the three natives is a Coral. Call them *I, II,* and *III*.

If *I* is a Coral, his name is Pedro and *II* is a Rose. *III* therefore is a Salmon, lying the first time when he says his name is Pedro. Therefore his second answer (*I* is Rodrigo) is true. This is absurd, so *I* is not a Coral.

Nor can *II* be a Coral, for his third answer must be truthful, and he says the Coral is *III*.

So *III* is a Coral; *I* therefore is a Rose; *II* is a Salmon. And *II*'s third answer is truthful. His first answer is truthful also.

24. david?

Let *A, B, C, D* stand for the four relevant names.

Then we can present the data according to the following scheme:

Owner	*A*	*B*	*C*	*D*		*m*	*n.*	*p*
Dog				*p*			*A*	
Cat	*m*	*p*				*n*		*C*

We have to determine *m, n,* and *p*.

1. *p* is either *A* or *B*. Let *p* be *A*. Then *n* is *D*, and *m* owns the cat *D*. Then *m* cannot be *A* (for *A* owns *C*), or *B* (for *B* would own *B*), or *C* (for *C* owns *A*), or *D* (for *D* would own *D*). Hence *p* is *B*. Now we have

Owner	*A*	*B*	*C*	*D*
Dog	*C*	*A*	*D*	*B*
Cat	*D*	*C*	*B*	*A*

So the dog David is owned by Charles.

25. the island of hu

Let *A, B,* and *C* stand for Smith, Jones, and Robinson respectively. Also, let S stand for "of the same race" and D for "of a different race." The only possible combinations for *AB, BC,* and *CA,* are S–S–S, S–D–D, D–S–D, D–D–S, and D–D–D. The following table gives the answers that will be given to the three questions: by a White—*W*; a Pink who answers the first question truthfully—*P*/1; a Pink who answers the second question truthfully—*P*/2; a Green who answers the first question truthfully—*G*/1; a Green who answers the third question truthfully—*G*/3; and a Green who answers no question truthfully—*G*/0, for each of these possible combinations:

Actual Relationship			W's answers	P/1's answers	P/2's answers	G/1's answers	G/3's answers	G/0's answers
AB	BC	CA	T T T	T L T	L T L	T L L	L L T	L L L
1 S	S	S	s s s	S D S	D S D	S D D	d d s	d d d
2 S	D	D	S D D	s s d	d d s	s s s	D S D	d s s
3 D	S	D	D S D	d d d	s s s	d d s	S D D	S D S
4 D	D	S	d d s	d s s	S D D	D S D	s s s	s s.d
5 D	D	D	d d d	D S D	S D S	d s s	s s d	s s s

The results indicated by capital letters are those which agree with the answers actually given. Hence the relationships between *A, B,* and *C* are those in line 1, which is clearly self-contradictory, or as in line 3. That is, *B* and *C* are of the same race (Green) and *A* is a White. The only way these facts will fit the answers as given is for Smith to be a White, Jones a Green (speaking truthfully in answer to the third question), and Robinson a Green (answering all three questions falsely).

26. artichoke, bergamot, and celery

1. Since Mr. Bergamot rode the horse he himself had trained, Mr. Artichoke rode the horse trained by Mr. Celery and Mr. Artichoke's horse is named Celery.

2. Placing our known data in tabular form gives

Mr.	*A*	*B*	*C*
owns	*C*		
trained	*x*	*y*	*z*
rode	*z*	*y*	*x*

3. This table can be completed under the terms of the problem in only one way:

Mr.	*A*	*B*	*C*
owns	*C*	*A*	*B*
trained	*B*	*C*	*A*
rode	*A*	*C*	*B*

4. Charlie Celery rode Bergamot.

27. the bank manager is?

Draw a circle to represent the plan of the table and number the seats, clockwise, 1 to 6; then 4 is opposite 1, 5 is opposite 2, and 6 is opposite 3.

Put the owner of the Chrysler at 1. Then Blue is at 2, and Red is at 6; the owner of the Buick is at 5, Green is at 4, and the owner of the Chevrolet (the stockbroker) is at 3. Since Red owns a Lincoln, not a Plymouth, the stockbroker is not Yellow and thus is either White or Black.

a. Suppose the stockbroker is White. Then Blue is the doctor. Black, opposite the engineer, cannot be at 5 and must be at 1, and Green is the engineer. Yellow is at 5, Red is the architect, and Blue is the owner of the Plymouth. Green must own the Ford; Black must be the lawyer, and Yellow the bank manager.

b. Suppose the stockbroker is Black. Then Red is the engineer. White, on the doctor's left, cannot be at 1 and must be at 5, and Green is the doctor. Yellow is at 1, Blue is the architect, and Green is the owner of the Plymouth. Blue must own the Ford; White must be the lawyer, and Yellow the bank manager.

It is impossible to tell without one further clue whether the correct identifications are those of (*a*) or of (*b*), but that is not the question; in either case, the bank manager's name was Yellow.

28. white and green

1. Leslie White is a female.

2. Miss Hilary Green is not betrothed to Mr. Sidney White and, since she cannot be betrothed to Miss Leslie White or to Hilary White, her prospective husband must be Mr. Alison White.

3. From the above it follows that Mr. Sidney White must be betrothed to Miss Alison Green.

4. Therefore Leslie Green is a man and, since he cannot be betrothed to Miss Leslie White or to Mr. Sidney White or to Mr. Alison White, he must be marrying Miss Hilary White.

29. strange!

1. Placing the known data in tabular form we have

Name	Charged with	Sentenced for
Burglary	x	
Dognaping	y	x
Counterfeiting		y
Y	z	
Z	counterfeiting	
Arson		embezzlement

2. Either Y or Z may be Arson.

3. If Y is Arson, Z cannot be Counterfeiting (Z is charged with counterfeiting), Arson (Y is), Embezzlement (for then Arson would be charged with, as well as sentenced for, embezzlement), or Dognaping (since Z is charged with counterfeiting and Dognaping is charged with arson). Z must then be Burglary and x becomes counterfeiting, giving us:

Name	Charged with	Sentenced for
Burglary	counterfeiting	
Dognaping	arson	counterfeiting
Counterfeiting		arson
Arson	burglary	embezzlement
Embezzlement		

4. The above table can only be completed with Burglary being sentenced for dognaping (the only possible alternative is that he be sentenced for burglary, which is inconsistent with the conditions of the problem).

5. If Z is Arson, Y cannot be Arson (Z is), Counterfeiting (for then both Arson and Dognaping would be charged with counterfeiting), Embezzlement (for then both Arson and Counterfeiting would have been sentenced for embezzlement), or Dognaping (this would make Dognaping charged with dognaping). Y must then be Burglary, giving us:

Name	Charged with	Sentenced for
Burglary	arson	
Dognaping	burglary	arson
Counterfeiting		burglary
Arson	counterfeiting	embezzlement
Embezzlement		

6. This table can only be completed by Burglary's being sentenced for dognaping (the only possible alternative is that he be sentenced for counterfeiting, which is not possible since it would result in Embezzlement being both charged with and convicted of dognaping).

7. Further trial will show that all other possible combinations lead to inconsistent results, and while we cannot—without more information from the Attorney General—completely solve this unprecedented case, we can state conclusively that Burglary was sentenced for dognaping.

part 2. reasoning

30. the light coin

You are given eight coins and told that one of them is not up to the standard weight.

If you had available a pair of balance scales, how would you proceed in order to identify the light coin in only two weighings?

31. how often does christmas fall on friday?

Given the Gregorian calendar (every ordinary year consists of 365 days; every year in which the year number is not a "century" year and is divisible by four is a "Leap Year" of 366 days—

February 29 being added—and every "century" year in which the century number is divisible by four is also a "Leap Year" of 366 days)—prove that the probability of Christmas falling on Friday is NOT one-seventh.

32. the lovesick cockroaches

The female cockroach at the southeast corner of a square is in love with the male cockroach at the northeast corner, who in turn is in love with the female cockroach at the northwest corner, who in turn is in love with the male cockroach at the southwest corner, who in turn is in love with the first cockroach (southeast corner). At a given instant all four cockroaches proceed on a pursuit curve directly toward their beloveds. Just what happens when they meet at the center of the square is not recorded. Assuming that they all travel at the same constant speed, how far (in terms of the length of the side of the square) does each cockroach travel before this meeting occurs?

33. chain smoker

When the poker party broke up at 2 A.M., Jerry Fullhouse rolled up his sleeves and got to work so that when his wife, who was out of town for the night, returned she would detect no signs of the evening's activities. The longer he worked, the more he felt the need of a smoke. He knew that all the cigarettes had been consumed long ago, but he also knew from experience that there was enough tobacco in four butts to make one adequate cigarette.

Carefully he gathered up the twenty-nine cigarette butts that had not yet been thrown in the open fire and, rolling and lighting his first makeshift cigarette, went back to work with renewed energy.

Before calling it a night and going to bed Jerry had made, and smoked, as many makeshift cigarettes as possible. How many butts did Jerry finally discard?

34. commuter service

The Deluxe Suburban Railroad provides excellent commuter service between Goodtown, a fine residential town, and Notsogood City (52.5 miles away), where most of the residents of Goodtown work. A train leaves Goodtown for the city at exactly 6 A.M. and every ten minutes thereafter. Upon reaching the city a train will discharge its passengers, turn around, take on any returning passengers and depart for Goodtown exactly 7.5 minutes after its arrival at the city. Upon reaching Goodtown a train will make a similar 7.5-minute layover before again departing for the city.

On the assumption that all trains maintain an average speed of 60 miles an hour, how many trains would Mr. Commuter pass in going from Goodtown to Notsogood City if he were riding in the train which left Goodtown at ten minutes past seven in the morning?

35. beefeater's convention

One hundred men attended the annual Beefeater's Convention. Of these seventy-three lived in Texas, seventy-eight lived in cities, sixty-eight lived in brick houses, and ninety-five drove their own cars.

We don't know how many of these men possess all four characteristics, but we know that at least x and not more than y live in brick houses in Texas cities and drive their own cars.

What is the value of x and y?

36. the walking fly

Although this problem appears in almost every collection of mathematical recreations, it is so interesting that we are including it for the benefit of those who may have been so unfortunate as not to have made its acquaintance previously. While it is a delightful problem, it is not particularly simple. There is an obvious answer, but the obvious answer is not necessarily the correct one.

A fly is in an oblong room that is thirty feet long, twelve feet wide, and twelve feet high. He is perched on an end wall one foot from the ceiling and six feet from each side wall. He wishes to proceed to a point on the other end wall one foot from the floor and six feet from each side wall.

In all problems involving flies it is customary to restrict the movements of the fly in some manner (See problem 6, section 2). This fly is not stupid, but for some reason or other he is a walking fly and not a flying fly.

If he takes the shortest path, how far must he walk to get to his destination?

37. chessboard

As anyone knows, a chessboard is an eight-by-eight square divided into sixty-four equal-size smaller squares.

Suppose you take a chessboard and remove any two squares from opposite corners. It does not matter which two you remove so long as they are opposites.

The problem is now to divide the remaining sixty-two squares into two-by-one oblongs. It looks easy enough, but after trying for a while you will always find that you wind up with two odd single squares left over.

There is a good reason for this: the problem is impossible of solution and there is a very simple commonsense proof. What is it?

38. u.s. coins

A certain article costs an amount which requires a minimum of four standard United States coins to pay for it. To purchase two of these articles would require a minimum of six coins. However, three of the article can be purchased for two coins.

What is the price of the article?

39. white hats

Two questions are asked in this problem. Many will recognize the first as an old friend, but most people will find the second question gives the problem a new aspect.

"The scholarship will be awarded," said the Dean to the three candidates—Tom, Dick, and Savvy—"to the winner in this little competition. I am going to blindfold you and then place a hat, which will either be green or white, on each of your heads. When I give the signal you will remove your blindfold and, if you see either of the other two is wearing a white hat, knock once on the table. As soon as you can tell me the color of the hat upon your own head, knock twice upon the table. Is this clear?" Assured that all three candidates fully understood the conditions, the Dean proceeded to blindfold them and then placed a white hat upon the head of each one.

At the word *Go!* all three removed their blindfolds and:

a. All knocked once practically simultaneously; after a slight pause, Savvy knocked twice and said, "My hat is white, Sir."

b. Savvy knocked three times so rapidly that it almost sounded like one knock and said, "My hat is white, Sir!"

How did Savvy know?

40. bigdome's park

Bigdome's Park has been left to the State Historical Society. This fine park is in the shape of a perfect circle with a diameter of five miles.

There are two monuments in the park. One reaches the older of these—which commemorates the Battle of Beancake—by entering the park at its northernmost point and proceeding due south for a mile. Now if, starting from this monument, one proceeds due west until one reaches the circumference of the park, then walks due south for a mile and a half, one reaches the second monument. This was built as a memento of Bigdome's first big coup on the stock exchange.

How far apart—as the crow flies—are the two monuments?

41. jim and judy

"Jim and Judy have been having a walking match," said Jon. "They agreed to do six miles each on the Westchester Road. Jim did the first stretch, from the twelfth to the sixth milestone, while Judy paced him on a bicycle. Then Jim took the bicycle while Judy walked from the sixth to the first milestone. They then went into the club to celebrate."

"Who won?"

"Judy won easily. Rather strange; because, if anything, her stretch of road was the more difficult. Do you think it can be that one walks more quickly after doing a few miles on a bicycle?"

Or is there another explanation?

42. jones was early

Commuter Jones caught an earlier train home than usual. His wife is one of those people who is always exactly on time, and she always arrives at the station just as Jones gets off his train.

Rather than wait at the station, Jones decided to walk toward home. He met his wife on her own way to the station. They reached home six minutes earlier than if he had waited at the station.

The car travels at a uniform speed, which is ten times Jones' speed on foot. (In addition to being punctual, Mrs. Jones obviously is a careful driver.)

Jones reached home at six. When would he have arrived if Mrs. Jones (warned of his change in plan) had met him at the station?

43. north

"What's happening in the card room?" I said to a friend at the club.

"Puffin and North are playing bridge against Hobo and Badluck. Puffin has just dealt four interesting hands. Puffin and North hold most of the red cards. Their red cards—let me think now—exceed the total of their opponents' red cards by the num-

ber of black cards in Hobo's hand. Puffin holds twice as many black cards as Hobo holds red ones."

"I don't see what I can deduce from that," I said.

"No? Not even if I tell you that Hobo has both the red Aces? He may easily make six No Trumps."

How many black cards does North hold?

44. the prince

"Each day," said the Wicked Fairy to the Prince, "you must fill so many sacks with acorns. You are to go on doing this until all the sacks are full.

"Moreover, your work will become increasingly strenuous. On each day after the first, you must fill double the number of sacks that you have so far filled. Thus, if you fill three sacks the first day, you must fill six the second day, eighteen the third day, and so on. Is that clear?"

"Perfectly clear," said the Prince, as he whistled up his Good Fairy. With her assistance the sacks practically filled themselves. At the end of seven days one-third of them were filled.

How long in all did the job take?

45. creaker *vs.* roadhog

In the early days of American motoring the Motor Parkway which ran for forty-five miles on Long Island was our only expressway and was the testing road for stock cars.

At the Motor Club in New York, Jalopy offered to bet that—given a five-minute head start—he could beat anyone to the end of the Parkway.

Newcar, who knew that Jalopy could not do better than forty miles per hour in his 1908 Creaker, and that his own new Roadhog could make sixty without even opening the throttle wide, eagerly accepted the bet.

What stratagem did Jalopy adopt to win his bet?

[Incidentally, this race actually took place around 1914—the winner was an old friend, the late Joseph B. Lee, of Brooklyn.]

46. chains

A farmer needs a chain fifteen links long and finds that he has available five pieces of chain of three links each. The blacksmith tells him that it will cost twenty cents to cut a link and thirty cents to weld a link previously cut. Assuming that the blacksmith does the job in the cheapest way, what will it cost the farmer to have the five pieces joined into one continuous chain of fifteen links?

47. triangles

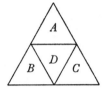

Each of the faces of a regular tetrahedron is divided as shown into four equilateral triangles. You are given four colors with which to paint this solid, and you must paint it in accordance with these rules:

1. Only two colors to be used on each face.
2. The same amount of each color paint is to be used.
3. No two triangles having one side in common to be painted the same color.

In how many different ways can you paint your tetrahedron?

48. as the tide rises

A small boat is moored to the dock. At noon the owner noted with satisfaction that the gunwale was exactly four feet above the water line. Three hours later the tide had risen exactly one foot and three inches.

How far will the gunwale be above the water line at 3 p.m.?

49. what time is it?

In cleaning out the top bureau drawer, Mrs. Adams found two wrist watches she and her husband had long ago discarded as impossibly balky timepieces. She decided to give them another trial. Therefore she wound them carefully and, after setting them accurately, started both watches going at the same time. Then she slipped her old watch on one wrist and her husband's on the other so that she might keep checking them at odd intervals during the day. An hour later she noticed that while her watch had gained one minute her husband's had lost two minutes.

Next morning when she looked at the watches again, it was seven o'clock by her old watch and six o'clock by her husband's. What time was it when she started the watches running?

50. the fresh-air fiend

"Why should I always travel by these stuffy closed buses?" thought the fresh-air fiend, Mr. Tan. "I know there are a few open buses about and I'll keep count of them the next time I go to see Smith."

The service concerned is that between South Town and North Town. The buses leave each end at 7 A.M. and then regularly every ten minutes thereafter. The journey takes ninety-five minutes and the buses wait five minutes at each end.

Mr. Tan lives near South North Circle, fifteen minutes from North Town; Mr. Smith lives at North South Circle, twenty minutes from South Town. (See sketch on next page.)

On his next trip to see Mr. Smith, Mr. Tan left South North Circle at 11:15 A.M. and arrived at North South Circle at 12:15 P.M. Counting the first bus he passed as he traveled toward North South Circle as number 1, he found that buses 3 and 12 were open buses and the others closed.

On his return trip, Mr. Tan left North South Circle at 5:40 P.M. and arrived back at South North Circle at 6:40 P.M. All the buses he passed on this return trip were closed except one group of four open buses in succession.

How many open buses are there and what are their numbers (assuming that the buses are numbered as mentioned previously)?

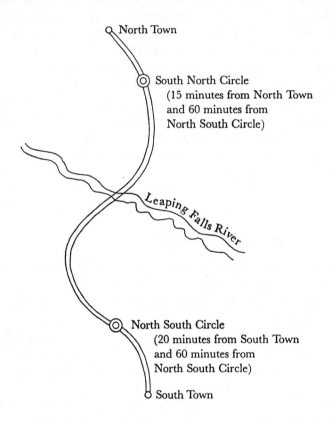

North Town

South North Circle
(15 minutes from North Town
and 60 minutes from
North South Circle)

Leaping Falls River

North South Circle
(20 minutes from South Town
and 60 minutes from
North South Circle)

South Town

51. mr. jones

The rich Mr. Jones, who was brought up in the county orphan's home, gives a Christmas party for the orphans every year and presents each of the boys and each of the girls a certain number of new dollar bills. Each year for the last three years he has increased the size of his gift to each boy and girl.

This particular year his total gift amounted to four hundred dollars and he gave each of the twenty-one girls eleven dollars. How many boys were there in the orphanage?

52. billiard plays?

Five men in our village are somewhat improbably named Billiard, Golf, Soccer, Chess, and Badminton. Each is an expert of one of the five games suggested by their several names. None is an expert in the game which his name suggests.

Each of the five, moreover, is engaged to the only sister of one of the others. In no case is the surname of a player's fiancée that which suggests the game in which he is expert. Nor is her surname that which suggests the game played expertly by the man whose name corresponds to the game which her fiancé plays. For example, if Soccer's game were badminton, and if Badminton's game were golf, Soccer would not be engaged to either Miss Badminton or to Miss Golf. Soccer is a chess expert. Chess is engaged to Miss Badminton. Golf is engaged to the sister of the billiards expert.

At what game is Billiard expert and what is the name of his fiancée?

53. much dithering

Messrs. Butcher, Carter, Farmer, Smith, and Wheelwright are all residents in the hamlet of Much Dithering.

They are the sons of a butcher, a carter, a farmer, a smith, and a wheelwright; they also pursue the vocations of butcher, carter, farmer, smith, and wheelwright. But none of the five bears the name of his father's—or of his own—vocation; and none follows his father's trade. None has as yet a son of his own.

Here are five clues, relating to these five people, with the aid of which this complex occupational tangle can quickly be resolved:

1. The wheelwright's son is engaged to a girl in the neighboring village of Long Suffering.

2. The village bridge four consists of Mr. Butcher, Mr. Wheelwright, the carter's son, and the smith.

3. Mr. Carter and the butcher's son married sisters.

4. The farmer's son and the wheelwright can frequently be

heard practicing duets, accompanied by Mr. Carter on his mandolin.

5. Of our five subjects, the smith is the only bachelor.

What is the occupation of each of the five villagers, and what is his father's?

54. the back-alley gang

"I shan't be satisfied," said the District Attorney, "till we've rounded up the Back-Alley Gang."

"Nor shall I," said Superintendent Gettum. "The first thing is to make sure who is who. I've just learned that Ezra Smith is either the Squealer, the Bruiser, or One-Eyed Mike."

"If he's not the Bruiser," said Inspector Brightone, "either Sam Wade or Solomon Isaacs is."

"Yes," said the D.A., "but Sam Wade might also be the Squealer or the Spider."

"Andy MacPhail," said Brightone, "is either the Sprinter or the Dixie Kid. But he can't be the Dixie Kid if Isaacs is the Bruiser."

"On the other hand," said Gettum, "Harry Bennett may be the Sprinter or he's Two-Gun Titus."

"Reverting for one moment to Isaacs," said the D.A., "who can he be if he's not the Bruiser?"

"The Spider," said Brightone. "The most dangerous guy of the lot."

"That's right," said Gettum. "Or, again, he might be One-Eyed Mike."

"I've heard nothing yet," said the D.A., "about David Davies. Except that he can't be the Spider."

"And he can't be the Bruiser," said Brightone. "He might—though I doubt it—be the Dixie Kid or Two-Gun Titus. But one or other of those is Morgan Evans."

"And that's all we know?" asked the D.A.

"Not quite," said Gettum. He consulted his notebook. "Here are some more scraps of information. If Davies is One-Eyed Mike, Isaacs is not the Spider. If Davies is the Squealer, Ezra Smith is not One-Eyed Mike. And, if Ezra Smith is the Bruiser, the Spider can't be Sam Wade."

Given that the police information is correct, who is who?

55. the seven housemasters

"When I came to this school," said Glowering, the recently appointed Dean of Seven Gables Boys' School, "I thought life was going to be simple. I am cursed with a poor memory, as you know. Judge my delight when I found that the school has seven houses—Blenkinsop, Brown, Dickory, Hickory, Jones, Smith, and Snooks—and that the seven housemasters have those identical seven names."

"And your delight turned to ashes in your mouth," said I, "when you found that not one of them bears the same name as his house?"

Glowering stared at me. "How on earth did you know that?"

Said I, primly: "Pray continue your narrative."

"It's not a narrative," said Glowering, suiting the action to his name. "It's a problem. You have to discover who is the house-master of each house. And here"—he produced his notebook—"are the data:

"All the housemasters are married with the exception of Jones and Brown. Brown (the house) is in the charge of the master whose name is the same as the name of Mr. Hickory's house. The matron at Dickory is engaged to the housemaster. The house which has the same name as the housemaster of Smith (the house) is in the charge of the namesake of the house which is ruled by Mr. Dickory. This chap, by the way (the namesake of Mr. Dickory's house, I mean), is married. And his wife's sister is married to the housemaster of Hickory."

"Do I have to do it in my head?" I asked.

"If you like," said Glowering. "You wouldn't be the first. And —I say—I've overlooked one clue. The three houses which have trisyllabic names are all in the charge of housemasters whose names are monosyllables."

What is the answer to Glowering's problem?

56. his instructor's age

A schoolboy sits for examinations in English, Algebra, History, Geography, and Chemistry. His instructors for these subjects are Mr. Andrews, Mr. Bailey, Mr. Chapman, Mr. Davies, and Mr. Elliott, not necessarily respectively, and they live at Helpem, Failem, Passem, Workem, and Sweatem, also not necessarily respectively. These towns are located, relative to each other and to the river, as shown in the sketch.

In each subject, 50 constitutes a pass. The boy's lowest mark is 45, and his highest mark is in Algebra; he fails in History, but obtains 64 in English and 56 in another subject. In Mr. Davies' subject his mark is one and one-half times that received in the subject for which the instructor who lives at Failem is responsible, and his average mark is 57.

Mr. Andrews, the youngest of the instructors, lives at Passem, and the instructor in whose subject the boy gets 56 lives on the side of the river opposite Mr. Andrews. Mr. Andrews' nearer neighbor on the same side of the river is younger than Mr. Davies; his nearer neighbor on the opposite side of the river is younger than Mr. Bailey, but older than Mr. Elliott. Mr. Chapman and the Chemistry instructor live on opposite sides of the river.

The ages of the History and Chemistry instructors are respectively the same as the boy's marks in Chemistry and History; and the ages of the other three instructors are respectively the same as one of the boy's marks, one of the boy's marks with the digits reversed, and the average age of the five instructors. No two instructors are the same age.

How old is the instructor who lives at Helpem?

30. the light coin

Strange as it may seem, the use of eight coins in this problem causes some individuals more difficulty than if nine coins had been specified. Place three coins on each side of the balance. If the scales balance, these six coins are normal and the light coin is one of the remaining two. Placing one of these on each side of the balance quickly locates the light coin. It is in the higher tray. On the other hand, if the first six coins did not balance, the defective coin is one of the three in the higher tray. If one of these three coins is placed on each side of the balance and it remains balanced, the light coin is the third. If it does not balance the light coin is the one in the higher tray.

This problem can be generalized if desired. 3^n coins can be handled in n weighings by the following method:

a. Place $3^{(n-1)}$ coins on each side of the balance and determine which group of $3^{(n-1)}$ coins contains the light coin.
b. Having determined which group of $3^{(n-1)}$ contains the light coin, subdivide it into three equal groups of $3^{(n-2)}$ each and place one of these on each side of the balance and thereby determine which group of $3^{(n-2)}$ contains the light coin.
c. By continuing this process the size of the group can be reduced by two-thirds at each successive weighing. Thus n weighings will be able to handle 3^n coins.

31. how often does christmas fall on friday?

The key to this problem lies in the fact that the Gregorian calendar, which is a cyclic calendar with a period of four hundred years, happens to consist of an exact number of weeks. The original Julian calendar, from which the Gregorian calendar was derived, was also a cyclic calendar—but its period was only four years. Every year consisted of 365 days except those years which were exactly divisible by 4 (that is, every fourth year) contained one extra day, or 366 days.

This was quite a simple calendar but it was not quite accurate and the day of the month began to creep out of the desired relation with the seasons. In the sixteenth century Pope Gregory devised the Gregorian calendar in which every year which was divisible by 4 was a leap year except the century years, which would be leap years only if the century number itself was divisible by 4 (for example, 1900 was not a leap year, but 2000 will be).

It is thus seen that the four-hundred-year cycle of the Gregorian calendar consists of 303 ordinary years of 365 days and only 97 leap years of 366 days. Since 365 days equal 52 weeks plus one day, the 303 ordinary years will contain 303 times 52 weeks plus 303 days. Similarly, the 97 leap years will contain 97 times 52 weeks plus 97 times 2 or 194 days (since there are, of course, two extra days in each of these years). This gives us for the four hundred years of the cycle 400 times 52 weeks plus $303 + 194 = 497$ days $= 71$ weeks, so that after four hundred years we are back where we started. For example, since Christmas fell on Sunday in 1960 it will also fall on Sunday in 2360, just as it fell on Sunday in 1560.

Since 400 is not divisible exactly by 7 it is obviously impossible for $\frac{1}{7}$ of the Christmases to fall on any given day of the week.

As a matter of information the following is the probability that Christmas will fall on any given day in the week when the year is selected at random (of course, for any given year the probability is either zero or one, that is, impossibility or certainty):

Day of the week	Number of times Christmas falls on any given day in any 400 years	Probability* of Christmas falling on any given day
Sunday	58	0.1450
Monday	56	0.1400
Tuesday	58	0.1450
Wednesday	57	0.1425
Thursday	57	0.1425
Friday	58	0.1450
Saturday	56	0.1400
Total	400	1.0000

*For comparative purposes $1/7 = 0.142,857, \ldots$

32. the lovesick cockroaches

While this problem can, with difficulty, be solved by calculus, it can be solved quite readily by simple reasoning. Since each cockroach starts at the same time and at all times heads directly toward his or her beloved, at any given instant the four will be located on the corners of a gradually decreasing (and rotating) square. Thus the path of any particular cockroach is at all times at right angles to the path of the one pursuing it. In other words the motion of the, say, northeast cockroach does not alter in any respect the distance the southeast cockroach has to travel in order to reach it. The length of the path each cockroach travels is, therefore, equal to the length of the side of the square.

33. chain smoker

The twenty-nine original butts yielded seven makeshift cigarettes and one butt. Then, after these seven cigarettes had been smoked there were $7 + 1 =$ eight butts. These in turn yielded two makeshift cigarettes. When Jerry had finished these two cigarettes he had only two butts to dispose of before going to bed.

34. commuter service

When the train in which Mr. Commuter is riding reaches the city it must have passed every other train making a similar run (since every train ahead of it had already reached the station and started back for Goodtown). The question of how many trains were passed thus becomes, in essence, a question of how many trains are necessary to maintain such a service.

Any particular train would reach the city 52.5 minutes after it had left Goodtown, depart for Goodtown 7.5 minutes later, arrive back at Goodtown 52.5 minutes later, and leave on its next trip for the city 7.5 minutes after that. The entire cycle therefore takes 120 minutes and, if there is to be a train every ten minutes, there must be twelve trains making the run; and, since any train must pass all the others on any given run, it follows that Mr. Commuter would pass eleven trains on his trip from Goodtown to Notsogood City.

35. beefeater's convention

Of the hundred men 73 lived in Texas; 78 lived in cities; 68 lived in brick houses; 95 drove their own cars.

This leaves 27 men who did not live in Texas; 22 who did not live in cities; 32 who did not live in brick houses; and 5 who did not drive their own cars. This totals 86. There are thus at most 86 men who lack one of these four characteristics and therefore at least 14 who must possess all four. Obviously, since only 68 lived in brick houses, and a greater number possessed the other characteristics, the maximum number that can have all four characteristics is 68.

Therefore at least 14 and not more than 68 of these 100 men live in Texas, in the city, in a brick house, and drive their own car. That is, $x = 14$ and $y = 68$.

36. the walking fly

The obvious answer is for the fly to either walk straight up to the ceiling, across the middle of the ceiling, and down the other end wall; or down to the floor, across the middle of the floor, and up the other end wall to his destination. Either of these paths is 42 feet (1 + 30 + 11 = 11 + 30 + 1 = 42).

In order to find a shorter path consider the room to be made of cardboard and cut a sufficient number of edges to permit the room to be unfolded as in the diagram (the room can be unfolded in various ways, of course, but trial will show that this one results in the shortest straight-line distance between the two points):

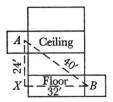

The fly must follow what is now a straight line from A to B. That is, he must travel diagonally up the end wall; diagonally across the ceiling; diagonally down the far wall; diagonally across the floor; and diagonally up the other end wall to his destination. (Note that this route, which is the shortest possible route, takes him over a portion of five out of the six sides of the room.)

The line AB is the hypotenuse of a right-angle triangle the sides of which are $AX = 24'$ and $BX = 32'$. This is the familiar 3–4–5 triangle and AB must equal $40'$, therefore, the walking fly must walk 40 feet.

An interesting variation of this problem is obtained by making the room twenty feet long, ten feet wide, and ten feet high. If the point of departure is on an end wall two feet from the ceiling and five feet from the side walls and the point of destination symmetrically located on the other end wall two feet from the floor and five feet from the side walls, the shortest path proves to be very little shorter than the obvious path (down to the floor, straight across, and then straight up—or the reverse). In fact they differ by less than .04 feet.

37. chessboard

The sixty-four chessboard squares are colored alternately white and black; any two-by-one oblong must consist of one black and one white square.

As you think about the board a little more you will see that opposite-corner squares are of the same color. You have removed two squares of the same color and when you try to make thirty-one pieces out of the rest of the board you will always wind up with two squares of the same color left over.

38. u.s. coins

A little quick figuring will show that the article costs seventeen cents.

> 17 cents—2 cents + 1 nickel + 1 dime = 4 coins
> 34 cents—4 cents + 1 nickel + 1 quarter = 6 coins
> 51 cents—1 cent + 1 half dollar = 2 coins

39. white hats

Considering first condition (*a*):

Since all three knocked promptly there had to be at least two white hats and, Savvy reasoned, "If I had a green hat on my head one of the other two (whom I know to be very intelligent) would quickly realize that the hat on his own head would have to be white and would promptly knock twice. Since neither has done so it follows that they are both looking at two white hats and, therefore, the hat on my own head cannot be green and must be white."

Considering condition (*b*):

During the time the Dean was describing the conditions of the test Savvy analyzed the situation as follows: "If the Dean places two hats of one color and one hat of the other color on our heads, the test will not be the same for each of us and will, therefore, not be a fair test. But I'm certain that the Dean will give us a fair test. It follows," reasoned Savvy, "that all the hats will be green or all will be white. I shall, therefore, knock twice the instant I see any green hat, or knock three times the instant I see any white hat."

40. bigdome's park

This problem practically solves itself *provided* you realize that after you have walked one mile south from the northernmost point of the park, due west until you hit the circumference of the park and then due south another mile and a half you will be on the east-west diameter of the park and that, taken along with the east-west and north-south diameters of the park, the last two legs of your walk form a perfect rectangle, one diagonal of which is a radius of the park (and hence equal in length to 2½ miles). Since the two monuments stand at the ends of the other diagonal of this rectangle, they also must be 2½ miles apart.

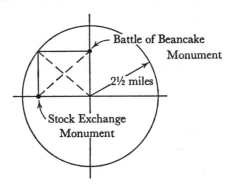

Battle of Beancake Monument

2½ miles

Stock Exchange Monument

41. jim and judy

A very simple one. From the twelfth to the sixth milestone is six miles, but from the sixth to the first is only five.

42. jones was early

Jones saved six minutes by walking part way. This meant that the car traveled three minutes less each way and hence was three minutes (at its uniform speed) from the station when Jones met it.

Since the car goes ten times as fast as Jones, it must have taken him thirty minutes to get to the meeting point.

But, if the car had been at the station to meet him, Jones would have gone the distance that he walked in three minutes instead of thirty and, therefore, would have arrived home twenty-seven minutes earlier than he actually did, or at 5:33 P.M.

43. north

At first the data seem inadequate. But let's see:

1. Since the red cards total 26, the difference in the numbers of red cards held by the two sides must be an even number. So Hobo holds an even number of black cards and, therefore, an odd number of red cards.

2. Hobo holds the two red aces. So he has three or five red cards (he cannot have seven or more cards because Puffin would then hold fourteen or more black cards, which is impossible).

3. If Hobo has three red cards, we can complete all four holdings as follows:

	Hobo	*Badluck*	*Puffin*	*North*
RED	3	5	7	11
BLACK	10	8	6	2

4. If Hobo has five red cards, he has eight black cards and Puffin will have ten black cards. Also, since the sum of Puffin's and North's red cards equals the sum of Hobo's and Badluck's red cards plus 8 (the number of black cards in Hobo's hand), Puffin and North hold seventeen red cards between them. However, since Puffin holds ten black cards, he can only hold three red ones. This leaves North holding fourteen red cards, which is impossible. Hence there is only one solution—that given in paragraph *3* above.

Therefore, North holds two black cards.

44. the prince

This is a variant of a fairly well-known "intelligence test." The *number* of sacks to be filled does not matter. The point is that each day the Prince is filling double the number so far filled; hence, when one-third are full, one more day's work will complete the job.

It, therefore, took the Prince eight days.

45. creaker *vs.* roadhog

Jalopy drove for three minutes and then pulled off the road and hid behind some bushes. After Newcar went by, Jalopy returned to the road and proceeded to the end of the Parkway at a leisurely pace. He passed Newcar and his Roadhog about ten miles from the finish.

It seems that when he didn't catch Jalopy after a reasonable time Newcar kept increasing his speed more and more until he burned out a bearing and could go no farther.

46. chains

Cut each link of one piece of chain and use the three cut links obtained thereby as connecting links to join the remaining four pieces of chain. Thus three cuts and three welds will do the job and more are not necessary. Since

$$3 \times 20 + 3 \times 30 = 150 \text{ cents} = \$1.50,$$

it will cost the farmer $1.50.

47. triangles

1. On each face, clearly, triangles *A*, *B*, and *C* must be painted one color and triangle *D* a second color.

2. Call the four colors *M*, *N*, *P*, and *Q*. Then each must be used for three triangles on one face. Two different arrangements, however, are possible. That is, if the tetrahedron stands on the base colored *M*, the other three faces in clockwise order can be painted *N–P–Q* or *N–Q–P*.

3. With each of these arrangements, nine variations of the second color of each face are possible. We can have

BASE		SIDE FACES	
MN	*NP*	*PQ*	*QM*
MN	*NQ*	*PM*	*QP*
MN	*NM*	*PQ*	*QP*

and other similar variations for *MP* and *MQ*.

So you can paint your tetrahedron in eighteen different ways.

48. as the tide rises

The same four feet. The boat rises and falls with the tide.

49. what time is it?

Obviously, the improbable assumption must be made that the watches continued to run at a constant rate. Under this assumption, every hour Mrs. Adams' watch will gain three minutes over her husband's watch, and it would take twenty hours for the two watches to be exactly one hour apart.

During this twenty hours Mrs. Adams' watch (which gains one minute an hour) must have gained twenty minutes and her husband's watch (which loses two minutes an hour) must have lost forty minutes. Hence the time is 6:40 A.M. It follows that it must have been 10:40 A.M. when she set the watches and started them going.

50. the fresh-air fiend

The complete cycle for any one bus is 95 + 5 + 95 + 5 = 200 minutes. To maintain the schedule of a bus every 10 minutes will, therefore, require $^{200}\!/_{10}$ = 20 buses.

In his morning trip Mr. Tan has seen every bus from the first bus to pass South North Circle after 11:15 A.M (the 10 A.M. bus from South Town arrives at South North Circle at 11:20 A.M., bus #1) to the last bus to pass North South Circle before 12:15 P.M. (the 11:50 A.M. bus from South Town arrives at North South Circle at 12:10 P.M.). He, therefore, passes twelve buses and of these #3 and #12 (the last bus he passed) were open buses.

In the evening Mr. Tan saw every bus from the first bus to pass North South Circle after 5:40 P.M. (the 4:30 P.M. bus from North Town arrives at North South Circle at 5:45 P.M.) to the last bus to pass South North Circle before 6:40 P.M. (the 6:20 P.M. bus from North Town arrives at South North Circle at 6:35 P.M.).

The bus which left North Town at 4:30 P.M. left South Town 100 minutes earlier, or at 2:50 P.M.; North Town at 1:10 P.M.; South Town at 11:30 A.M. and is, therefore, bus #10. In a similar manner it will be seen that the last bus he passed was bus #1.

It is thus evident that Mr. Tan has passed all the buses engaged in maintaining the service between South Town and North Town. Bus #12 is known to be an open bus. It is also known that the only open buses among buses #10 through #20 and #1 are four open buses in succession. Since buses #10 and #11 are closed, it follows that buses #12, #13, #14, and #15 must be open.

There are thus five open buses on the service and their numbers are 3, 12, 13, 14, and 15.

51. mr. jones

Since he gave $231 to the girls there was $169 left over for the boys. There was evidently more than one boy and each boy got the same amount, which must be more than $2. But the only factors of 169 are 1, 13, and 169. Therefore, there must have been 13 boys, each receiving $13.

52. billiard plays?

1. The given data can be written as follows:

Player	Game	Fiancée
Billiard		
Golf		Miss X
Soccer	chess	
Chess		Miss Badminton
Badminton		
X	billiards	

While the relationships given below are excluded:

——	p	Miss Q
P	q	

2. Now Mr. X cannot be Billiard (since he plays billiards), Golf (his fiancée is X) Soccer (his game is chess), or Badminton (Miss Badminton is engaged to Chess). Hence Mr. X is Mr. Chess.

3. This gives:

Player	Game	Fiancée
Billiard		
Golf		Miss Chess
Soccer	chess	
Chess	billiards	Miss Badminton
Badminton		

4. Mr. Soccer cannot be engaged to Miss Chess or Miss Badminton (they are already engaged to someone else), or Miss Billiard (because Soccer's game is chess and Mr. Chess plays billiards). Therefore he must be engaged to Miss Golf. It follows that Mr. Billiard must be engaged to Miss Soccer and Mr. Badminton to Miss Billiard.

5. Again from the problem conditions it is seen that Mr. Golf cannot play soccer since Mr. Soccer plays chess, and this is inconsistent with Mr. Golf's being engaged to Miss Chess. Since he cannot be expert in chess (Soccer is), or Golf (suggested by his own name), or billiards (Chess is), he must be expert in badminton. It then follows that Mr. Badminton plays soccer (Billiard cannot—he is engaged to Miss Soccer) and, by elimination, Mr. Billiard plays golf.

6. So Mr. Billiard is the golf expert and is engaged to Miss Soccer.

53. much dithering

1. Note first that the wheelwright's son is not married and is the smith—(*1*) and (*5*).

2. Mr. C is not the butcher's son (*3*), wheelwright's son (*1* above), or the farmer's son (*4*); *i.e.*, he is the smith's son. Also he is not·the wheelwright (*4*).

3. Mr. B is not the carter's son (*2*), the wheelwright's son [(*2*) and *1* above], or the smith's son (Mr. C); *i.e.*, he is the farmer's son and therefore cannot be the farmer.

4. The farmer's son is not the wheelwright (*4*), so Mr. *B* is the carter.

5. The wheelwright's son is the smith (*1* above). Therefore he must be Mr. *F*. Also, Mr. *W* is not the carter's son (*2*).

6. The rest is now immediately deducible:

Mr. Butcher (farmer's son) is the carter. Mr. Carter (smith's son) is the butcher. Mr. Farmer (wheelwright's son) is the smith. Mr. Smith (carter's son) is the wheelwright. Mr. Wheelwright (butcher's son) is the farmer.

54. the back-alley gang

Call Smith, Wade, Isaacs, etc., *S, W, I, D, M, B, E*.

Call the Squealer *1*, the Bruiser *2*, One-Eyed Mike *3*, the Spider *4*, the Sprinter *5*, the Dixie Kid *6*, and Two-Gun Titus *7*.

We then have these data:

1. *S* is *1, 2,* or *3*.
2. *W* is *1, 2,* or *4*.
3. *I* is *2, 3,* or *4*.
4. *M* is *5* or *6*. ⎫ Hence *M, B,* and *E* must be
5. *B* is *5* or *7*. ⎬ *5, 6,* and *7* in some order.
6. *E* is *6* or *7*. ⎭
7. *D* is neither *2* nor *4* (and is therefore *1* or *3*).
8. If *M* is *6, I* is not *2*.
9. If *D* is *3, I* is not *4*.
10. If *D* is *1, S* is not *3*.
11. If *S* is *2, W* is not *4*.

Now consider the group *S, W, I, D*. On the basis of data 1, 2, 3, 7 we have these possibilities:

	A	B	C	D	E	F
S can be *1, 2, 3*	1	1	2	2	3	3
W can be *1, 2, 4*	2	4	1	4	2	4
I can be *2, 3, 4*	4	2	4	3	4	2
D can be *1, 3*	3	3	3	1	1	1

But of these, A and C are ruled out by 9; D by 11; and E and F by 10. Hence, B is the only solution which conforms to the data. Now consider 4, 5, 6. These are the only possibilities:

M	5	6
B	7	5
E	6	7

But the second one is ruled out by 8.

Hence Ezra Smith is the Squealer; Sam Wade is the Spider; Solomon Isaacs is the Bruiser; David Davies is One-Eyed Mike; Andy MacPhail is the Sprinter; Harry Bennett is Two-Gun Titus; Morgan Evans is the Dixie Kid.

55. the seven housemasters

1. Mr. Dickory is not the housemaster of Blenkinsop, Dickory, or Hickory (trisyllabic names), Brown or Jones (since the namesake of his house is married), or Smith (Smith—the house → Mr. X; x—the house → Mr. Y; y—the house → Mr. Dickory). Hence he is the housemaster of Snooks.

2. The housemaster of Hickory is not Mr. Blenkinsop, Mr. Dickory, or Mr. Hickory (trisyllabic names), Mr. Brown or Mr. Jones (he is married), or Mr. Snooks (the sister of the namesake of Mr. Dickory's house is married to the housemaster of Hickory); therefore he must be Mr. Smith.

3. Mr. Hickory is not housemaster of Blenkinsop, Dickory, or Hickory (trisyllabic names), Brown (the namesake of Mr. Hickory's house is housemaster of Brown), Snooks (*1* above), or Smith (for this would make Mr. Smith housemaster of Brown, contrary to *2* above). Mr. Hickory must thus be housemaster of Jones.

4. Since Mr. Hickory is housemaster of Jones it follows that Mr. Jones is housemaster of Brown.

5. Mr. Blenkinsop is not housemaster of Blenkinsop, Dickory, or Hickory (trisyllabic names), or Snooks, Jones, or Brown (*1*, *3*, and *4* above). Mr. Blenkinsop is, therefore, housemaster of Smith.

6. This leaves Mr. Snooks as the housemaster of Blenkinsop

and Mr. Brown as the housemaster of Dickory (who must be a bachelor).

7. To summarize:

Smith → Mr. Blenkinsop;   Blenkinsop → Mr. Snooks;
Snooks → Mr. Dickory;   Dickory → Mr. Brown;
Brown → Mr. Jones;   Jones → Mr. Hickory;
Hickory → Mr. Smith.

56. his instructor's age

Let A, B, C, D, and E represent Mr. Andrews, Mr. Bailey, Mr. Chapman, Mr. Davies, and Mr. Elliott, respectively.

1. From the second paragraph—
   a. D does not live at Failem.
   b. The mark in D's subject is divisible by 3. It cannot therefore be 64 or 56; nor can it be 45, the lowest mark.
   c. The mark in the Failem instructor's subject is divisible by 2, so it is not 45; nor is it 64 or 56, for if it were, the mark in D's subject would be 96 or 84, either of which would put the average above 57.
   d. Therefore the mark in D's subject and mark in the Failem instructor's subject are respectively

   $\frac{3}{5}$ and $\frac{2}{5}$ of $[5 \times 57 - (64 + 56 + 45)]$,

   that is, 72 and 48. It follows that D is the Algebra instructor and that the Algebra grade is 72; and that the Failem instructor, who teaches neither Algebra nor English, assigned a mark of 48.

2. From the third paragraph—
   a. A lives at Passem.
   b. The mark in the Helpem instructor's subject is 56 (since the instructor in whose subject the boy's marks are 56 lives on the opposite side of the river from A and he cannot live at Failem, because that instructor assigned a mark of 48),

so that he does not teach Algebra, English, or History; D therefore does not live at Helpem.

c. D does not live at Workem (A's nearest neighbor on the same side of the river lives at Workem and is younger than D).

d. Therefore D lives at Sweatem (the other towns are all eliminated in paragraphs 1(d), 2(a), 2(b), and 2(c) above).

e. The instructor who lives at Failem is not B or E (A's nearest neighbor on the opposite side of the river lives at Failem and is younger than B and older than E).

f. Therefore C lives at Failem.

g. The Helpem instructor is not the Chemistry instructor (C and the Chemistry instructor live on opposite sides of the river).

h. It follows that the Geography instructor lives at Helpem, and 56 is the Geography mark.

i. C is not the Geography instructor [2(f) and (h)], the Chemistry instructor (C and the Chemistry instructor live on opposite sides of the river), the English instructor (the student received 64 in English and 48 from the instructor who resides at Failem), or the Algebra instructor [1(d) above]. Thus we have the fact that C is the History instructor, and 48 is the History mark.

j. The Chemistry mark is 45 (all other marks are accounted for).

3. From the fourth paragraph—

a. The ages of C (the History instructor) and the Chemistry instructor are respectively forty-five and forty-eight.

b. A and E are both younger than C and thus cannot teach Chemistry. Therefore B is the Chemistry instructor.

c. It follows at once that:

B lives at Workem
A is the English master, and
E lives at Helpem.

4. The order of ages is   *D*
                          B (forty-eight)
                          C (forty-five)
                          E
                          A

a. Therefore *E*'s age is the average of the five ages.
b. The marks are 72, 64, 56, 48, and 45, and the reversed marks are 27, 46, 65, 84, 54.
c. *A*'s age and *D*'s age must be chosen, one from each of these sets of numbers, so that *A*'s age is less than forty-five and *D*'s greater than forty-eight. It follows that *A*'s age is twenty-seven (the only age in the two series that is less than forty-five) and *D*'s age is fifty-six, sixty-four, or seventy-two.
d. If *D*'s age were seventy-two, *E*'s age would be $(72 + 48 + 45 + 27)/4 =$ forty-eight, and if *D*'s age were sixty-four, *E*'s age would be $(64 + 48 + 45 + 27)/4 =$ forty-six, but *E*'s age is less than forty-five, therefore, *D*'s age is fifty-six (he cannot be forty-five or forty-eight) and *E*'s age is $(56 + 48 + 45 + 27)/4 =$ forty-four.

We see, therefore, that Mr. Elliott, who lives at Helpem, is forty-four years of age.

5 the answers are whole numbers

Many of the problems in this section fall within the category of Diophantine Equations, named for the great mathematician Diophantus of Alexandria, who studied and wrote about problems of this nature around A.D. 300. (A Diophantine Equation is an equation with two or more unknowns in which the solutions are limited to whole numbers.)

There is no reason to turn away from these problems just because the subject may be new to you. Problems of this kind are omitted from most conventional courses in algebra because usually they do not possess a unique solution, but they are not difficult to solve.

Take a sample problem: "Twice Jody's age in years plus Sandra's age in years equals seven. How old are Jody and Sandra?"

This statement does not give enough data for a solution since it yields only the equation $2x + y = 7$. There are two unknowns and only one equation, so mathematically an infinite number of answers are possible. However, from the conditions of the prob-

lem we know that x and y are both positive integers and therefore the only possible answers are:

Jody is 3 and Sandra 1
Jody is 2 and Sandra 3
Jody is 1 and Sandra 5

We don't have a single solution; but if we were given the
additional information that Jody is older than Sandra only the
answer "Jody is 3 and Sandra 1" would be correct, and we would
have a Diophantine Equation with only one solution.

Another very simple Diophantine Equation is present in the
catch question: "I have two coins that are worth fifty-five cents
between them. One is not a half-dollar. What are they?"

Here we are restricted to very specific integral values of x and
y for our two coins, and eventually we see that one is a nickel
(not a half-dollar) and the other *is* a half-dollar.

The general form of a first-degree Diophantine Equation with
two unknowns is $ax + by = c$, where a, b, and c are integers and
x and y are unknowns.

Except in the case in which a and b have a common factor that
is not a factor of c, there will be an infinite number of integral
solutions to each such equation. In this exceptional case there will
not be any integral solution. This can be proved as follows:

Let n be a factor of a and b that is not a factor of c. Then, if
you divide the equation by n, you have

$$\frac{ax}{n} + \frac{by}{n} = \frac{c}{n}.$$

If x is an integer and a/n is an integer, ax/n is an integer.
Similarly, by/n will be an integer if y is an integer; but, since n is
not a factor of c, c/n cannot be an integer. Therefore both x and
y cannot be integers.

There is a direct mathematical approach to the integral solutions of this equation. However, this approach is not essential for
the Diophantine Equations to which the problems in this section
will lead. You can solve them by trial and error. Here is a sample equation and the trial-and-error solution.

$17x = 39y + 8$, which can also be written

$$17x = 34y + 5y + 8.$$

Dividing by 17 gives $x = 2y + \dfrac{5y + 8}{17}$.

Since x and y will both be integers, then $(5y + 8)/17$ must be an integer.

Give y the successive values 0, 1, 2, 3, and so forth; $5y + 8$ becomes 8, 13, 18, 23, 28, 33, 38, 43, 48, 53, 58, 63, 68, and so forth. It isn't necessary to go any farther because 68 is exactly divisible by 17. So, when $y = 12$ we have $x = 28$ as the smallest possible positive integral solution to the equation. Also, each increase of 17 in the value of y will give another integral value to the fraction and will represent an increase of 39 in the value of x. Other positive integral solutions are $x = 67, y = 29; x = 106, y = 46$; and so on up. Also, by decreasing y and x we obtain negative integral solutions of $x = -11$, $y = -5$; $x = -50$, $y = -22$; and so on.

The most famous Diophantine problem of all is Fermat's Last Theorem, in which he states that for all integral values of n that are greater than 2 there do not exist integers a, b, and c, all different from zero, so that the following equation is true:

$$a^n + b^n = c^n.$$

Fermat stated that he had a proof of this but never gave it and so far no one has been able either to prove or disprove it. However, it has been proved true for all values of n up to 4002.

When $n = 2$ the equation becomes $a^2 + b^2 = c^2$.

This is, of course, the Pythagorean Theorem (in a right-angle triangle the square of the hypotenuse is equal to the sum of the squares of the other two sides). There are an infinite number of integral solutions to this, starting with $a = 3$, $b = 4$, and $c = 5$, which is the triangle used by the Egyptians in surveying the Nile lowlands. If a, b, and c are multiplied by the same number the proportion remains the same, so that 6, 8, 10; 9, 12, 15; etc., are also right triangles. The next right triangle of different shape is

5, 12, 13. If you want a simple method for finding all possible integral right triangles let

$$c \text{ (the hypotenuse)} = k(m^2 + n^2)$$
$$a \text{ (one side)} \quad\quad = k(m^2 - n^2)$$

and $\quad\quad\quad b \text{ (the other side)} \; = k(2mn)$

where k, m, and n are any three positive integers with m greater than n. For example:

If $k = 1$, $m = 2$, and $n = 1$	then	$a = 3$, $b = 4$, $c = 5$.
If $k = 1$, $m = 3$, and $n = 2$	then	$a = 5$, $b = 12$, $c = 13$.
If $k = 1$, $m = 4$, and $n = 1$	then	$a = 15$, $b = 8$, $c = 17$.
If $k = 1$, $m = 3$, and $n = 1$	then	$a = 8$, $b = 6$, $c = 10$,

which is, of course, just double the 3, 4, 5 triangle. Also, if

$$k = 3, m = 2, \text{ and } n = 1 \quad\quad \text{then} \quad\quad a = 9, b = 12, c = 15,$$

which is, of course, just three times the 3, 4, 5 triangle.

1. the farmers' market

At the Farmers' Market Mary purchased 100 pieces of fruit for $8. If apples cost five cents, peaches seven cents, and pears nine cents, how many more pears than apples did Mary buy?

2. jody and sandra

Jody and Sandra enjoyed playing little mathematical games. Jody wrote down a number and asked Sandra to guess it. "How about a little information"? asked Sandra. "The number is exactly twice the product of its two digits," replied Jody.

What was the number?

3. cashier's error

Just before closing time at the corner bank, a man presented a check to the cashier to have it cashed. Obviously in a hurry to finish his day's work, the cashier accidentally transposed the dollars and cents, thus giving the man in dollars the amount shown on the check as cents and in cents the amount marked as dollars. On the way home the man spent $3.50. When he counted his money after this purchase, he found to his surprise that he now had twice the amount the check had called for.

What was the original value of the check?

4. the crossed ladders

There is an old-favorite algebra problem in which you have a fairly narrow street with buildings on each side with a ladder extending from the base of one building to the building directly opposite and another ladder extending from the base of this second building back to the first building. You are given the lengths of the two ladders (which are unequal) and the height of their intersection from the line connecting the lower end of the two ladders (which are, of course, at the same level) and asked to find the width of the street. This problem usually results in a very awkward fourth-degree equation with no integral solutions. As a variant of this problem consider the following:

Determine a set of values for the quantities involved in the problem so that they will all be an exact number of inches. That is, the quantities *AB, AC, AD, AE, AF, BC, BE, CD,* and *EF* are all integers. The lengths of the two ladders must not be equal; this would make the problem trivial.

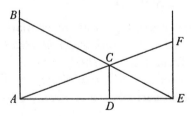

5. egg money

A farmer on his way to town to buy eight bags of chicken feed (costing two dollars a bag) noted with interest that the hundred silver coins his wife had given to him from her egg money to pay for the feed consisted of dimes, half-dollars, and silver dollars but contained no quarters.

How many dimes, half-dollars, and silver dollars did his wife give him?

6. wendy, sally, and debby

A few years ago Wendy, Sally, and Debby were greatly excited when they received their Christmas presents from Grandpa and found that he had given each one a five-dollar bill for each year of her age.

After squeals of joy as Grandpa announced he intended to follow the same formula thereafter, the more mathematical of the three noted that she had received just as much as the other two together.

Several years later one of the girls noticed that this time she had received exactly half the total given the other two.

Last Christmas, when one of the girls was twenty-one, it was noticed that the number of years that had elapsed since their first gifts (under the five-dollar-a-year-of-age formula) was two-thirds the number of five-dollar bills distributed that first year.

How old were the other two last Christmas?

7. monkey business

Hanging over a pulley is a rope with a weight at one end; at the other end hangs a monkey of equal weight. The rope weighs eight ounces per foot. The sum of the age of the monkey and his mother is eight years. The mother is as many years old as the weight of the monkey in pounds. The mother is twice as old as the monkey was when the mother was half as old as the monkey will be when the monkey is three times as old as its mother was when she was three times as old as the monkey. The weight of the rope and weight is half again as much as the combined weight of the weight and the monkey.

How long is the rope?

8. scholarships

At the annual meeting of the school board it was proposed that the extra five thousand dollars available be awarded to the nineteen most deserving boys and girls in the district—each boy and girl to receive an exact number of dollars. All the boys were to receive the same amount. This was also the case with the girls. However, each boy was to receive $150 more than each girl.

While there was some objection to the extra amount the boys received, the motion was carried.

How much did each boy and girl receive?

9. birthday cakes

Joan, Jane, and John are each celebrating birthdays today, and it was discovered that, although when John is half as old as Jane will be when he is half as old as Joan is now, Jane will be twice as old as Joan was when John was half as old as he is now. Jane is now older than John by the number of years represented by the product of the difference in years between Joan and herself and one-half the difference in years between the ages of Joan and John.

How many candles are there altogether on the three birthday cakes?

10. which is nearer?

It is a thousand miles from New York to St. Louis. A train leaves New York for St. Louis at eighty miles an hour. Two hours later a train leaves St. Louis for New York at sixty miles an hour. Which is nearer St. Louis when they meet?

11. muggins

"How much money did you say I would get under my grandfather's will?" asked Muggins of Calculus, the trustee.

"Square your grandfather's age [in years]," said Calculus.

"Divide the figure you get by your own age. Square your own age. Divide the figure you get by your grandfather's age. Add the two quotients together and subtract their sum from a thousand. Multiply the result by ten. That's the number of dollars you will get."

Muggins, who is a bit slow in the uptake, replied, "I square my grandfather's age and divide the figure I get by my own age. I square my own age and divide the figure I get by my grandfather's age. I subtract this quotient from the first quotient and multiply the result I get by ten. That's the number of dollars I get."

Calculus laughed. "Have it your own way," he said, "the result is the same thing either way."

How old is Muggins?

12. mary

"How old are your children?" I asked Mrs. Cosine, merely by way of making conversation.

To my horror she produced some notes from her handbag. "You're Mr. Askem, aren't you?" she said. "Miss Black, who is Isobel's principal, suggested I should give you this problem."

And this is what I read:

"The children's combined ages (in years) total three times the age (in years) of Sandra, who is three years younger than Isobel. The cube of Isobel's age, plus the square of Sandra's age, plus Mary's age (all in years) total an exact number of hundreds.

"How old is Mary?"

13. knights vs. bishops

Representatives of Knight Town and Bishopsburg met to play chess on one hundred boards. Both men and women players took part, each town producing more than fifty male players.

More women played from Bishopsburg than from Knight Town.

It was arranged that, so far as possible, men should be matched against men and women against women. This meant that there were only mixed matches (men against women) to the extent to which the number of men players representing Knight Town exceeded the number of men representing Bishopsburg.

On the boards at which men were matched against men, Knight Town won three matches out of five. On the boards at which women were matched against women, Bishopsburg won two matches out of three. On the boards where Knight Town's surplus men met Bishopsburg's surplus women, the Knight Town players won two-thirds of the matches. No game ended in a draw. The result was a narrow win for Knight Town by fifty-one games to forty-nine.

How many men played for Knight Town?

14. the price went down

Stopping at the apple stand for his weekly purchase of fifty cents' worth of apples, Professor Digit was pleased to find that he was given five more apples than usual.

"Hmm!" he said. "I see the price has gone down ten cents per dozen."

"What was the new price per dozen?

15. four brothers and their gin game

Gus, Bernie, Ralph, and Buddy played gin rummy together.

1. Gus, the family expert, beat each of his brothers by a number of points equal to the brothers' age in years.

2. The youngest brother beat each brother except Gus by a number of points equal to Gus's age in years.

3. Gus won 170 points.

4. Bernie beat Buddy by 6 points and wound up 6 points less loser than Buddy.

5. Ralph won 86 points.

6. When the youngest brother was born, Gus's age was twice that of all the others together.

How old is each brother?

16. the ancient order of the greens

Extract from a report of the Games Subcommittee of the Ancient Order of the Greens:

"We recommend the purchase of 2 billiard tables and 12 dartboards. A poll of our members shows that 17 per cent would like to play billiards, 22 per cent snooker, and 28 per cent darts. These percentages include 137 members who have said they would like to play all three games. No one expressed a wish to play two."

There was some grumbling by 1462 Greens who did not want to play any of the three games.

How many Greens are there in all?

17. mrs. overtwenty

"I always answer my daughter's questions truthfully," said Mrs. Overtwenty smugly. "Fibs, however well-intentioned, must react unfavorably on a child's upbringing."

"What would you say if she asked you how old you are?" asked Mabel. "Most mothers, I've noticed, are inclined to be vague about that."

"She did ask me that, only yesterday," said Mrs. Overtwenty. "She also asked me what my own mother's age was. I gave her a truthful answer, though not an intelligible one. 'Darling,' I said, 'if you squared your granny's age (in years) and also squared my age (in years) and subtracted the second square from the first, the difference would be 1817.'"

"And what did darling say to that?"

"Ooh, you are old, Mummy, aren't you? Much older than Booboo!"

How old is Mrs. Overtwenty?

18. how many men?

One of the young stable hands at the Bar Z Ranch had a mischievous sense of humor. One afternoon as he returned to the stables from the corral, a groom asked him how many men there were tending the horses in the corral.

"What I saw was eighty-two feet and twenty-six heads," the stable hand replied saucily; then he neatly slipped out of the way, leaving the groom to figure out his own answer.
How many men were tending the horses?

19. seventeen pencils

A man bought seventeen pencils for seventy-two cents. He paid one cent more for each red pencil than for each plain pencil.
How many of each kind did he buy at what prices?

20. jones' children

Jones' will distributed $48,000 among his surviving children. Some of his children died between the time the will was drawn and the time of Jones' death, with the result that each of the survivors received $11,200 more than if Jones had died immediately after drawing up the will.
How many children did Jones have and how many predeceased him?

21. mrs. quigley

Mrs. Quigley was a straightforward person in all respects but one. Whenever the question of her age came up, she invariably resorted to this kind of indirect answer: "My husband is my senior, you know, and his age is the very reverse of mine. The difference between our ages is one-eleventh of their total. That should make things clear enough, it seems to me."
How old are Mr. and Mrs. Quigley?

22. the *rhind papyrus*

One of the problems contained in the *Rhind Papyrus* (a practical handbook of Egyptian mathematics written about 1700 B.C.) may be stated as, "Divide one hundred loaves among five men in such a way that the shares shall be in arithmetic progression and that one-seventh of the sum of the largest three shares shall be equal to the sum of the smallest two." (It is believed that the Egyptians solved this and similar problems by the method of false position.)

If the Egyptians had asked, "What is the least number of loaves there could have been that would permit each of the five to receive a whole number of loaves when divided in accordance with the above instructions?" what would you have answered?

23. poppy

"Poppy," asked his mathematically inclined granddaughter, "please give me a farm problem about when you were a boy?"

"All right," said Poppy. "When I was a boy my father had six pastures. On each of these pastures there grazed the same number of animals. He sold all these animals, which consisted of cows, hogs, and sheep, to seven dealers. Each dealer bought the same number of animals, paying eighteen dollars for a cow, five dollars for a hog, and three dollars for each sheep. The total amount the dealers paid my father was $451.

"What was the largest number of animals my father could have had and how many of each kind were there?"

24. the graustark cabinet

The Graustark Cabinet voted on a number of issues.

The Prime Minister reported to the King: "On every issue the Government has a clear majority.

"I was the only Minister who voted with the majority on each of the issues raised."

"What was the attitude of the other Ministers?" asked the King.

The Prime Minister consulted his notes. "The Minister of War," he said, "on eight occasions voted with the minority. The Lord Chancellor on seven occasions. The Postmaster and the Minister for Air on five occasions each. The Home Secretary, the Governor of the Bank, and the Attorney General on three occasions each. And every other member of the Cabinet on one occasion only."

"Odd," said the King.

"I can tell Your Majesty something odder still," said the Prime Minister. "On each issue every member of the Cabinet voted; the majority was not the same on any two issues; and we achieved every possible distribution of votes, as between the majority and the minority."

How many Ministers are there in the Graustark Cabinet?

25. the two cousins

Steven was quite excited at the news that his two cousins from Quebec would be visiting him that summer because he had never seen them.

"Are they old enough for me to play with?" he asked his father.

"Figure it out for yourself," his father baited him. "They are not old enough to vote and if you add the square of Roger's age to George's age, the sum is 183. Now don't bother me any more."

How old are the visiting cousins?

26. professor algebra's garden

The other day Professor Algebra had several friends in to see his garden, of which he is justly proud. In a rather small space he had developed several new strains of tomatoes and at the same time had grown a large variety of flowers of such quality that they had been awarded many blue ribbons.

As would be expected, everything about Professor Algebra's garden has an integral length in feet. The garden proper is a square surrounded by a paved walk on each side. These four walks are all of different widths (all widths of the walks and sides of the gardens proper are, as said before, an exact number of feet). There is a sundial in the middle of the narrowest walk, the next wider walk is one foot wider, the next still another foot wider, and the widest is still one more foot wider. The entire lot, gardens and walks included, is again a perfect square with sides an exact number of feet in length.

If the area of the entire square lot is 621 more square feet than the area of the square garden, how wide is the strip of walk on the side opposite the sundial?

27. coconuts

Four men are shipwrecked on an island on which there is nothing to eat except coconuts. They soon observed that the monkeys were eating large numbers of coconuts so, to protect their food supply, they gathered all the coconuts into a pile. That night, after they had gone to bed, one of the four awoke and said to himself, "I don't know these other three men, they may not be honest; I'm going to take my share of the coconuts now." He therefore divided the coconuts into four equal piles, found three over (which he threw to the monkeys), hid his fourth and put the remaining three-fourths back into one pile and went to bed. A second man woke up and, going through the same mental process, divided the remaining coconuts that he found into four equal piles, found three over (which he threw to the monkeys), hid his fourth and, putting the remaining three-fourths back into one pile, went back to bed. This process was repeated by the other two in turn. The next morning the pile was obviously smaller than it had been the night before but, as all four men had a guilty feeling, none said anything and readily agreed when it was suggested that each one take his share and use it as he saw fit. Accordingly they divided the remaining coconuts into four equal piles (this time there were none left over for the monkeys) and each took his quarter of the remainder.

What was the least number of coconuts that there could have been initially?

28. a box of chocolates

"Grandpa," said young Evelyn, "had such a bright idea the other day. He said he wanted to find out how greedy we were."

"What happened?" I asked.

"You tell him, Charles," said Evelyn.

"Well," said Charles, "he produced a big box of chocolates and gave it to Evelyn——"

"No, he gave it to Kitty," corrected Evelyn.

"Anyway," said Charles, "he gave it to one of them. Then he explained that all of us could eat as many chocolates as we liked. At the end of the day, Kitty would report the number of chocolates eaten. Then each of us would receive, from Grandpa, one penny for each chocolate eaten by somebody else, less three cents for each chocolate eaten by himself (or herself)."

"What a cunning scheme!" said I. "I'll bet it didn't cost Grandpa very much."

"Didn't it just!" said Clare. "It cost him all of $4.69."

"Whoever ate all the chocolates?" I asked.

"Edna ate the most," said Charles. "Then came David. I forget about the rest of us. But I remember that no two of us ate the same number, and that each of us ate at least one."

How many children in all ate chocolates, and how many did Edna eat?

29. "seventeeners"

"Rather fun, Miss Scribble's intelligence tests," said Jane. " 'Seventeeners,' she called them. We had a lot of them the two weeks we were in camp."

"Why 'Seventeeners'?" I asked.

"For some unknown reason," replied Jane, "the winner of each test collected seventeen cents. But that isn't all; each of us was fined a cent for each test she didn't win."

"At that rate," said I, "some of you must have wound up in the red."

"We did," said Jane. "I, for instance, lost thirty cents. Still, none of us complained. Miss Scribble, who made up the difference between our winnings and our losses, contributed $3.60. If we'd had twelve tests a day, as was originally planned, she'd have had to pay out even more. It was a curious affair though, in that each of us won at least one test but no two of us won the same number."

How many tests did Jane win?

solutions

1. the farmers' market

This problem becomes a quickie if you note that Mary spent $1 more than the seven-cent average. Since each apple is two cents under this average and each pear is two cents over, Mary must have purchased fifty more pears than apples.

Here is the mathematical solution of this problem. Let

x = the number of apples Mary purchased,
y = the number of peaches Mary purchased,
z = the number of pears Mary purchased.

Then $$x + y + z = 100$$
and $$5x + 7y + 9z = 800.$$

If y is eliminated by subtracting seven times the first equation from the second equation the result is

$$-2x + 2z = 100$$
or $$z = 50 + x$$

from which it follows that Mary purchased fifty more pears than apples.

2. jody and sandra

Let the number be xy, with x the digit in the tens place and y the digit in the units place; then

$$10x + y = 2x(y)$$
or $$y = (2y - 10)x$$
or $$x = \frac{y}{2y - 10}$$

where x and y are integers.

Obviously, y must be greater than 5. This therefore gives

y	x
6	$\frac{6}{2} = 3$
7	$\frac{7}{4} = 1\frac{3}{4}$
8	$\frac{8}{6} = 1\frac{1}{3}$
9	$\frac{9}{8} = 1\frac{1}{8}$

y must have been 6 and x have been 3.

Jody must have selected the number 36.

3. cashier's error

There are two variables involved in this problem: d, representing the number of dollars in the check, and c, representing the number of cents. Since there is information given only for a single equation, this problem technically has an indeterminate solution.

The very nature of the problem makes fractional values impossible either for the dollars or for the cents. Also, both of these numbers must be less than 100, or there could be no transposition. Making the calculations in terms of cents, we can state the value of the check originally as $100d + c$. But the teller, transposing the figures, gave the man $100c + d$. It is from this that the one equation in the problem is derived:

$$100c + d - 350 = 2(100d + c).$$

This linear diophantic equation is readily solved as follows:

Reducing the equation above to the form of an improper fraction gives

$$c = 2d + 4 + \frac{3d - 42}{98}.$$

But

$$\frac{3d - 42}{98}$$

must equal an integer, say e; then

$$e = \frac{3d - 42}{98} \quad \text{or} \quad d = \frac{98e + 42}{3} = 32e + 14 + \frac{2e}{3}$$

and for this to be an integer e must equal $3k$ (where k is any integer), making

$$d = 98k + 14$$
and $$c = 199k + 32.$$

But, since both c and d must be less than 100, k must equal zero, thus making $c = 32$ and $d = 14$. The check was for \$14.32.

4. the crossed ladders

Since the problem calls for all the quantities involved to be integral it is necessary to have the right-angle triangles ACD and CDE of a form that allows all sides to be rational numbers. The simplest rational relationship is the famous 3–4–5 ratio. For the length of the ladders to be different the line CD must be the shorter leg of one of the triangles and the longer leg of the other. Hence line CD must be a multiple of 3 times 4 if it is to be integral. If you start with CD equal to 12 inches the values will be

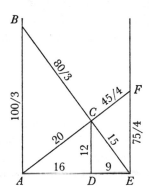

Obviously, if all values are multiplied by 12 the result will be that all will be integers and give:

The height of the cross point, CD, equal to 144 inches.
The length of the long ladder, BE, equal to 500 inches.
The length of the short ladder, AC, equal to 375 inches.
The width of the street, AE, equal to 300 inches.

Naturally this method can be generalized. All that is necessary is to let the line *CD* equal a multiple of *a* times *b* (where *a* and *b* are the legs of an integral right-angle triangle) and multiply all values by the appropriate number to clear of fractions. It is quickly seen that this value is also the product *ab*. This will give the following general solution:

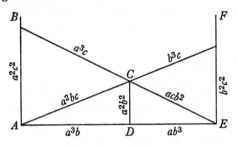

The length of one ladder $= bc(a^2 + b^2) = bc^3$
The length of the other ladder $= ac(a^2 + b^2) = ac^3$
The width of the street $= ab(a^2 + b^2) = abc^2$

If a, b, and c are integer solutions of the equation $a^2 + b^2 = c^2$ all the quantities will be integers. If $a = 4$, $b = 3$, and $c = 5$ the numerical result will be the same as that derived directly in the original solution.

5. egg money

Let $x =$ the number of dimes
                $y =$ the number of half-dollars
                $z =$ the number of silver dollars.
Then $x + y + z =$ the total number of coins $= 100$
and

$10x + 50y + 100z =$ the total cost of the feed in cents $= 1600$

Subtracting 10 times the first equation from the second gives:
$$40y + 90z = 600$$
or $$4y + 9z = 60$$

(y and z are both positive integers).

This equation can, of course, readily be solved by the method explained in the introduction to this section but, since it is quite a simple one, it can also be solved rather quickly by trial and error. The only values of z that need checking are 1 to 6 inclusive, since higher values would result in y being a minus quantity. These values give

z	y
1	$5\frac{1}{4} = 12\frac{3}{4}$
2	$4\frac{2}{4} = 10\frac{1}{2}$
3	$3\frac{3}{4} = 8\frac{1}{4}$
4	$2\frac{4}{4} = 6$
5	$1\frac{5}{4} = 3\frac{3}{4}$
6	$\frac{6}{4} = 1\frac{1}{2}$

The only positive integral value of z which results in y being a positive integer is 4. Therefore, $z = 4$; $y = 6$; $x = 90$.

The farmer's wife gave him ninety dimes, six half-dollars, and four silver dollars.

6. wendy, sally, and debby

This problem looks very complicated but simplifies quickly if you notice that the second condition (one girl's age was half the sum of the other two) is true all the time as long as it is true at any time.

That is, if their ages are a, b, and c and $a + b = 2c$ then k years earlier or later $a + k + b + k$ will equal $2(c + k)$ as a simple identity.

If their ages at the time of the first Christmas under the new formula were x, y, and $x + y$, it follows that $2x$ must equal $y + x + y$ and x will equal $2y$.

This will make y the youngest and their ages on the first occasion y, $2y$, and $3y$. The total of these three ages is $6y$ and the number of years that elapsed between the first gift and the time when one girl was twenty-one must be two-thirds this amount, or $4y$ years. Their ages last Christmas would have been $5y$, $6y$, and $7y$ years.

Since one girl's age at this occasion was twenty-one and y is an integer, the only possible value for y is 3.

Last Christmas the young ladies were fifteen, eighteen, and twenty-one, respectively.

7. monkey business

Let m = the age of the monkey.

Let x = the age of the monkey when his mother was three times as old. Then $3x$ = the age of the mother at that time. The difference in ages of the two is thus $3x - x = 2x$ and the present age of the mother is $m + 2x$.

The monkey will be $3(3x)$ when he is three times as old as his mother was when she was three times as old as the monkey. If the mother was half this old she would be $\frac{1}{2}(3)(3x)$. The monkey at this time would be $\frac{1}{2}(3)(3x) - 2x$. Since the mother is twice this old

$$m + 2x = 2[\frac{1}{2}(3)(3x) - 2x] = 9x - 4x = 5x$$

or
$$m = 3x,$$

and since the sum of their ages is eight years

$$m + (m + 2x) = 8.$$

Combining these two equations gives

$$3x + (3x + 2x) = 8$$

or
$$x = 1$$

and
$$m = 3 \quad \text{and} \quad m + 2x = 5.$$

Thus the weight of the weight = the weight of the monkey = the age of the mother in years = 5 pounds and, if R equals the weight of the rope

$$R + 5 = \frac{3}{2}(5 + 5) = 15$$

or
$$R = 10$$

and, since each pound is equivalent to two feet, the rope must be twenty feet long.

8. scholarships

Let g represent the number of girls. Then $19 - g$ will be the number of boys. In addition, let x be the number of dollars each girl receives. This will make the number of dollars each boy receives equal to $x + 150$. Since the total value of all the scholarships is $5000, it follows that

$$gx + (19 - g)(x + 150) = 5000$$

where g and x are integers.

Simplifying the above equation gives

$$gx + 19x + 2850 - gx - 150g = 5000$$
$$19x - 150g = 2150$$

or

$$x = \frac{150g + 2150}{19} = 7g + 113 + \frac{17g + 3}{19}.$$

Since x and g are integers, $(17g + 3)/19$ must also be an integer, say z.

Then

$$z = \frac{17g + 3}{19}$$

or

$$g = \frac{19z - 3}{17} = z + \frac{2z - 3}{17}$$

and here again $(2z - 3)/17$ must be an integer. This will be the case if we let $z = 17w - 7$. Then

$$g = 17w - 7 + \frac{34w - 14 - 3}{17} = 19w - 8$$

and

$$x = 133w - 56 + 113 + \frac{17(19w - 8) + 3}{19}$$

$$= 133w + 57 + 17w - 7 = 150w + 50.$$

That is to say

$$g = 19w - 8$$
$$x = 150w + 50$$

where w is any integer. However, if w is zero or less, the number of girls is negative—which is not possible under the conditions of the problem. Similarly, if w is greater than 1 the number of girls is greater than 19—which is also impossible. Therefore w must equal 1. It follows that

$$g = 19 - 8 = 11 \quad \text{and} \quad x = 150 + 50 = 200.$$

Each of the eleven girls received $200 and each of the $19 - 11 = $ eight boys received $200 + $150 = $350.

9. birthday cakes

1. From ". . . when he is half as old as Joan is now," and "Jane is now older than John . . . ," John is clearly the youngest. Let x equal the difference in years between Jane and Joan. Let y equal the difference in years between the ages of Joan and John. Let z equal John's age.

2. From the first part of the problem

a. When John is half as old as Joan is now, Jane will be

$$\text{Jane} + \frac{\text{Joan}}{2} - \text{John} = K_1.$$

b. When John is $\tfrac{1}{2}K_1$ Jane will be twice

$$\left(\text{Joan} - \frac{\text{John}}{2}\right),$$

therefore

$$\frac{\text{Jane} + \dfrac{\text{Joan}}{2} - \text{John}}{2} - \text{John} = 2\left(\text{Joan} - \frac{\text{John}}{2}\right) - \text{Jane}.$$

c. This can be simplified to

$$6x - y = 3z$$

where $x = $ Jane's age $-$ Joan's age
              $y = $ Joan's age $-$ John's age
and $z = $ John's age.

3. From the second part of the problem

$$\frac{xy}{2} = x + y.$$

a. From the very nature of the problem, the only answers to this equation which need be considered are those in which x and y are whole numbers. The only solutions to this equation in integers are

$x =$	-2	3	4	6
$y =$	1	6	4	3

(Note that y cannot be minus, since we know that Joan is older than John.)

b. Substituting these values in the equation $6x - y = 3z$, we have

$x =$	-2	3	4	6
$y =$	1	6	4	3
$z =$	$-1\frac{2}{3}$	4	$2\frac{2}{3}$	11
Joan's age $=$		10		14
Jane's age $=$		13		20

c. Obviously $x = -2$ and $x = 4$ do not fit the other conditions because this results in John's age being an improper fraction, contrary to the conditions of the problem.

d. Also John cannot be eleven years old and Joan fourteen, for then John is already more than half Joan's age, contrary to the conditions of the problem.

e. It follows that John is four, Joan is ten, and Jane is thirteen, and that the total number of candles is $4 + 10 + 13 = 27$.

10. which is nearer?

Neither. They are in the same place.

If you insist on making a serious problem of this and find out where they meet, the answer is 640 miles from New York.

In two hours the first train has gone 160 miles; they are 840 miles apart when the second train starts. Their speed of approach is 60 + 80 or 140 miles per hour, so it takes six hours more before they meet. The train from New York will have been traveling for eight hours and will have gone 640 miles.

11. muggins

Let $\qquad$ $x =$ Muggins' grandfather's age
$y =$ Muggins' age.

Then $\qquad 10\left(1000 - \left(\dfrac{x^2}{y} + \dfrac{y^2}{x}\right)\right) = 10\left(\dfrac{x^2}{y} - \dfrac{y^2}{x}\right)$

where x and y are integers. This equation simplifies to

$$1000 = \frac{2x^2}{y} \qquad \text{or} \qquad \sqrt{5y} = \frac{x}{10}.$$

Since the answers to this equation must be integers, y must equal 5, 20, 45, or higher, which makes the values for x 50, 100, 150, or more.

Moreover, since Muggins is obviously an adult, the value of $y = 5$ must be eliminated. Likewise the fact that it is unreasonable for the grandfather to be 150 or more leads to the conclusion that Muggins must be twenty and his grandfather one hundred.

12. mary

Let Isobel's age be x. Then Sandra's age is $(x - 3)$ and, if y is Mary's age, we have

$$x + (x - 3) + y = 3(x - 3)$$

and $x^3 + (x - 3)^2 + y =$ an exact number of hundreds $= 100c$ where c is a positive integer.

From the first of these equations

$$y = 3x - 9 - x - x + 3 = x - 6,$$

and substituting this value of y in the second equation gives

$$x^3 + x^2 - 6x + 9 + x - 6 = 100c$$
or $$x^3 + x^2 - 5x + 3 = 100c$$
or $$(x - 1)(x - 1)(x + 3) = 100c.$$

Since the right-hand side of this equation is an even number it follows that x must be an odd number.

This equation will be true for any value of x so that

$$(x - 1) = 10k$$

where k is any positive integer. This in turn is true for $x = 11$, 21, 31, and so forth (incidentally the smallest value of x that would make $[x + 3]$ a factor of 100, and which also makes the equation a true one, is 47). It follows that Isobel is eleven, Sandra eight, and Mary five. Higher values of x, Isobel's age, such as, twenty-one (which fit the equation) are ruled out by the reference to Isobel's principal. At the age of twenty-one or more it is to be presumed that Isobel will be married or away at college.

13. knights vs. bishops

Suppose that m men played for Knight Town and n men played for Bishopsburg. Then we have

$$\frac{3n}{5} + \frac{100 - m}{3} + \frac{2m - 2n}{3} = 51$$

or $$9n + 500 - 5m + 10m - 10n = 765$$

or $$5m - n = 265$$

where $100 > m > n > 50$ and m and n are integers. For this to be true n obviously must be a multiple of 5, say

$$n = 5k,$$

then $$5m - 5k = 265 \quad \text{or} \quad m = k + 53.$$

It follows that k must be some integer from 11 to 19 inclusive (otherwise n will be equal to or less than 50 or equal to or greater than 100). But if k is 14 or more, n is greater than m, which is not possible. So k must be 11, 12, or 13.

Notice also that $100 - m = 100 - k - 53 = 47 - k$ must be divisible by 3. Since this is not true when k equals 12 or 13 and is true when k equals 11, the only possible solution which fits all conditions is the one corresponding to $k = 11$, or

$$m = 64 \quad \text{and} \quad n = 55$$

and there must have been sixty-four men playing for Knight Town.

14. the price went down

Let x equal the new price in cents per dozen apples. Then $(x + 10)$ cents is the original price.

At the original price the professor would have received 12 times $50/(x + 10)$ apples. At the new price he received 12 times $50/x$ apples. But the latter figure is 5 greater than the former, or

$$\frac{12 \times 50}{x + 10} + 5 = \frac{12 \times 50}{x}.$$

Multiplying both sides of this equation by $x(x + 10)$ gives

$$600x + 5x^2 + 50x = 600x + 6000$$
or $\qquad 5x^2 + 50x - 6000 = 0$
or $\qquad x^2 + 10x - 1200 = 0$
or $\qquad (x + 40)(x - 30) = 0$
or $\qquad x = 30 \text{ or } -40$

Under the conditions of the problem x cannot be negative, so the new price of apples was thirty cents a dozen.

15. four brothers and their gin game

This is one of the class of problems which is fairly straightforward if approached from the proper direction and practically impossible of any solution other than trial and error if not.

The point is that it is necessary to determine by reasoning which is the youngest brother before setting up the equations. Obviously, Gus is the oldest brother (statement 6). Also, Gus and Ralph are winners and Bernie and Buddy are losers (3, 4, and 5). Since the youngest brother won an amount corresponding to twice Gus's age and lost only an amount equal to his own age, he must have been one of the two winners and so is Ralph.

Let G = Gus's age; B = Bernie's age; R = Ralph's age; and b = Buddy's age. From facts 1 and 3 you get

$$B + R + b = 170 \qquad (1)$$

and from facts 2 and 5 you get

$$2G - R = 86. \qquad (2)$$

From facts 1, 2, and 4 you get

Bernie's losses = $B + G - 6$
and   Buddy's losses = $b + G + 6$.

But from fact 4 Bernie's losses are 6 points less than Buddy's losses, hence

$$B + G - 6 = 6 + b + G - 6$$
or $$6 + b = B. \qquad (3)$$

Also, when Ralph was born Gus would have been $G - R$ years old; Bernie would have been $B - R$ years old; Buddy would have been $b - R$ years old; and Ralph zero years old. Fact 6 then gives the final equation needed

$$G - R = 2(B - R + b - R + 0)$$
or $$G = 2B + 2b - 3R. \qquad (4)$$

Substituting the value of G from equation 4 in equation 2 gives

$$4B + 4b - 6R - R = 86$$
or $$4B + 4b - 7R = 86. \qquad (5)$$

Substituting the value of B from equation 3 in equations 1 and 5 gives

$$6 + b + R + b = 170 \qquad \text{or} \qquad 2b + R = 164 \qquad (6)$$
$$24 + 4b + 4b - 7R = 86 \qquad \text{or} \qquad 8b - 7R = 62. \qquad (7)$$

Subtracting equation 7 from four times equation 6 gives

$$11R = 594 \quad \text{or} \quad R = 54.$$

Substituting this value of R in equation 6 gives

$$2b + 54 = 164 \quad \text{or} \quad b = 55$$

and substituting this value of b in equation 3 gives

$$6 + 55 = B \quad \text{or} \quad B = 61.$$

Finally, substituting the value of R in equation 2 gives

$$2G - 54 = 86 \quad \text{or} \quad G = 70.$$

Gus is seventy; Bernie is sixty-one; Buddy is fifty-five; and Ralph is fifty-four.

16. the ancient order of the greens

There are much harder ways to solve this problem than the one given below.

Let $x =$ the number of Greens who like only billiards,
$y =$ the number of Greens who like only snooker,
$z =$ the number of Greens who like only darts,
$T =$ the total number of Greens.

Since no Green expressed a liking for only two of the games, if the number of Greens who liked all three games, 137, is added to the number who liked only billiards, x, the result will be the total number of Greens who like billiards and, since this number is 17 per cent of the total number of Greens, T, it follows that

$$x + 137 = 0.17T.$$

A similar line of reasoning leads to the equations

$$y + 137 = 0.22T$$
$$z + 137 = 0.28T.$$

To get the fourth equation needed note that the total number of Greens must equal the sum of those who like one game, those who like three games, and those who like none of them, or

$$x + y + z + 137 + 1462 = T.$$

Subtracting the sum of the first three equations from the last equation gives

$$1462 - 2 \times 137 = T - 0.67T = 0.33T$$

or $$0.33T = 1188$$

or $$T = 3600.$$

There were 3600 Greens in all.

17. mrs. overtwenty

Let Mrs. Overtwenty's age be x and her mother's age be y. Then, since Mrs. Overtwenty's age in years squared is 1817 less than her mother's age in years squared it follows that

$$y^2 - x^2 = 1817.$$

But, since the difference between two squares equals the product of the sum and difference of the quantities involved, this can be rewritten as

$$(y + x)(y - x) = 1817.$$

1817 can be factored as follows

$$1817 = 1 \times 1817$$

or $$1817 = 23 \times 79.$$

As 23 and 79 are both prime numbers, no other possibilities exist for factoring 1817 into two separate integral factors. Since the first possibility is ruled out under the conditions of the problem, we have

$$y + x = 79$$

and $$y - x = 23.$$

Adding these two equations gives

$$2y = 102 \quad \text{or} \quad y = 51$$

and subtracting the second from the first gives

$$2x = 56 \quad \text{or} \quad x = 28.$$

And Mrs. Overtwenty, who had no intention of telling her curious friend how old she was, has done so inadvertently. Mrs. Overtwenty is all of twenty-eight years old.

18. how many men?

Obviously, the young stable hand was counting both horses and men.

Let x equal the number of men and y the number of horses. Then there would be $2x$ plus $4y$ feet and $x + y$ heads, or

$$2x + 4y = 82$$
and
$$x + y = 26.$$

Subtracting twice the second equation from the first gives

$$2y = 30$$
or
$$y = 15$$
and
$$x = 11.$$

There must have been eleven men and fifteen horses.

A more elegant approach is this one:

If all twenty-six heads belonged to men there would be only fifty-two feet. This leaves $82 - 52 = 30$ feet to be accounted for by the horses. Each horse having two more feet than a man, this would mean that there had to be fifteen horses and $26 - 15 =$ eleven men, as noted in the solution above.

19. seventeen pencils

Let x be the number of pencils bought at y cents each. Then $17 - x$ must be the number of pencils bought at $y + 1$ cents each. In that case

$$xy + (17 - x)(y + 1) = 72$$
$$xy + 17y - xy + 17 - x = 72$$
$$17y = 55 + x$$
$$y = 3 + \frac{4 + x}{17}.$$

The only value for x less than 17 that makes y an integer is $x = 13$, whereupon $y = 4$.

The man bought thirteen plain pencils at four cents each and four red pencils at five cents each.

20. jones' children

This problem is easily solved by constructing the following table:

Number of survivors	Each share	Total
2	$24,000	$48,000
3	16,000	
4	12,000	
5	9,600	
6	8,000	
8	6,000	
10	4,800	
12	4,000	

In order for a share to have increased by $11,200, it would have to be either $16,000 or $12,000. By inspection we see that if there had been ten children at the time the will was drawn and three had survived, each share would have increased by $11,200.

There is also an answer that must be dismissed because of its absurdity. Jones could have left $800 to each of sixty children, and if only four had survived the shares would be $12,000 each or $11,200 more than the original share.

There were therefore originally ten children and of these ten, seven predeceased Jones.

21. mrs. quigley

Mrs. Quigley is forty-five and her husband is fifty-four.

If the digits of Mrs. Quigley's age are represented as AB, their mathematical value would be $10A + B$; her husband's age, accordingly, would have the value $10B + A$. These lead to the equation

$$(10B + A) - (10A + B) = \tfrac{1}{11}([10A + B] + [10B + A])$$

or

$$B = \frac{10A}{8}.$$

This equation, indeterminate since it involves two unknowns, can easily be resolved under the conditions of this problem. Both A

and B, being digits, must be integers ranging from 1 to 9. To clear the equation, A must be 8 or 4. If it is 8, then $B = 10$, which is impossible here. Hence A is 4 and B is 5.

22. the *rhind papyrus*

Let

$$
\begin{aligned}
x + 4d &= \text{largest share,} \\
x + 3d &= \text{next largest share,} \\
x + 2d &= \text{next largest share,} \\
x + d &= \text{next largest share,} \\
x &= \text{smallest share. Then} \\
5x + 10d &= z \quad \text{and} \\
\tfrac{1}{7}(3x + 9d) &= (2x + d)
\end{aligned}
$$

where z is the total number of loaves. Simplifying:

$$3x + 9d = 14x + 7d \quad \text{or} \quad 2d = 11x \quad \text{or} \quad 10d = 55x$$

and thus

$$5x + 55x = z$$

or

$$60x = z.$$

If we let $x = 1$ the total number of loaves will be 60, a whole number, but we will have

$$d = 5\tfrac{1}{2}.$$

In order to have this common difference a whole number it will be necessary for x to be an even number. We then have as the smallest number of loaves which will meet our requirements $x = 2$ and $d = 11$. This makes the five shares

2, 13, 24, 35, 46—for a total of 120.

As a check note that

$$\tfrac{1}{7}(24 + 35 + 46) = \tfrac{1}{7}(105) = 15 = 2 + 13,$$

as required.

23. poppy

Since there were six pastures and the number of animals in each pasture was the same, the total number of animals must have been $6r$, where r is a positive integer. Also, since each of the seven dealers bought the same number of animals, this quantity must also be divisible by 7. Hence $7k = r$, where k is a positive integer, and the total number of animals is $42k$.

Since the total price was $451 and the cheapest animal cost $3, the greatest possible number of animals is $45\frac{1}{3}$ or 150. Therefore

$$42k < 150.$$

From this it follows that the maximum number of animals that there could have been is obtained when $k = 3$, or the maximum possible number of animals is 126.

Let x equal the number of cows, y equal the number of hogs, and z equal the number of sheep. Then

$$x + y + z = 126$$
and
$$18x + 5y + 3z = 451.$$

Subtracting three times the first equation from the second gives

$$15x + 2y = 73$$
or
$$2y = 73 - 15x$$
or
$$y = 36 - 7x + \frac{1 - x}{2}$$

and if $x = 2h + 1$ then $y = 36 - 7(2h + 1) - h$
or
$$y = 29 - 15h.$$

Since x is minus if $h < 0$ and y is minus if $h > 2$ it follows that the only possible values of h are 1 and 0. Hence $x = 1, y = 29$, and $z = 96$, or $x = 3, y = 14$, and $z = 109$.

But the animals were "cows, etc." It follows that the first possibility is eliminated and the largest number of animals Poppy's father could have had was 126—3 cows, 14 hogs, and 109 sheep.

24. the graustark cabinet

Assume that there are n members in the Cabinet. Then the total number of *noes* voted is thirty-four for the eight specified members plus one more for each member over eight, or

$$A = 34 + (n - 8) = 26 + n.$$

Since, (a) on every issue every member of the cabinet voted, (b) the majority was not the same on any two issues, and (c) every possible distribution of votes was obtained, there must have been one issue on which the voting was each of the following possibilities:

Number in favor	Number against
n	0
$n - 1$	1
$n - 2$	2
$n - 3$	3
$n - 4$	4
—	—
—	—
—	—
$n/2 + 1$	$n/2 - 1$

if the number of members is even and $(n + 1)/2$ $(n - 1)/2$ if the number of members is odd. In both cases the number of *noes* form an arithmetic progression with a total of

$$S_{\text{odd}} = \frac{(n - 1)/2}{2}[(n - 1)/2 + 1] = \frac{n^2 - 1}{8}$$

when n is odd and

$$S_{\text{even}} = \frac{n/2}{2}(n/2 - 1) = (n/2 - 1)n/4$$

when n is even. But S must equal A. Trying the case where n is even gives

$$26 + n = (n/2 - 1)n/4$$

or $\qquad n^2 - 10n - 208 = 0.$

Since this equation has no integral solutions, it follows that n cannot be even. When n is odd

$$26 + n = (n^2 - 1)/8$$

or $\qquad n^2 - 8n - 209 = 0$

or $\qquad (n - 19)(n + 11) = 0$

or $\qquad n = 19 \text{ or } -11.$

Since n negative has no meaning in this problem, it is evident that there must be nineteen members in the Graustark Cabinet.

25. the two cousins

Let x equal Roger's age and y equal George's age. Then

$$x^2 + y = 183$$
or
$$y = 183 - x^2.$$

At first glance it would appear that there was an infinite number of solutions, but y must be positive, therefore x cannot be greater than 13. Also y cannot be greater than 21 (since neither cousin is old enough to vote) and hence (since $183 - 12^2 = 183 - 144 = 39$) x cannot be 12 or less. It follows that x must equal 13 and

$$y = 183 - 13^2$$
or
$$y = 183 - 169 = 14.$$

Roger is thirteen and George fourteen.

26. professor algebra's garden

Let the length of one side of the lot be x and the length of one side of the garden be y; then, from the last paragraph, we know that

$$x^2 - y^2 = 621.$$

Factoring both sides gives

$$(x + y)(x - y) = 3 \times 3 \times 3 \times 23.$$

Since x and y are integers $(x + y)$ and $(x - y)$ must also be integers. This can only be true if

	Case I	$(x - y) = 1$	and	$(x + y) = 621$
or	Case II	$(x - y) = 3$	and	$(x + y) = 207$
or	Case III	$(x - y) = 9$	and	$(x + y) = 69$
or	Case IV	$(x - y) = 23$	and	$(x + y) = 27$.

Solving these pairs of equations gives

Case I	$x = 311$	and	$y = 310$
Case II	$x = 105$	and	$y = 102$
Case III	$x = 39$	and	$y = 30$
Case IV	$x = 25$	and	$y = 2.$

Cases I and II are eliminated because there will not be room enough to put walks of the type prescribed by the problem around such a large garden in a lot only slightly larger.

Case IV is eliminated since Professor Algebra could not possibly have grown all he did in a space 2 feet by 2 feet. It follows that the lot is 39 by 39 feet and that the garden is 30 by 30 feet.

From the fact that the lot and the garden are both square it follows that the narrowest width of walk, w, will be opposite the widest side, $w + 3$. Also the side $w + 1$ feet in width will be opposite the side which is $w + 2$ feet in width, or

$$x - y = w + w + 3 = w + 1 + w + 2 = 2w + 3$$

or $39 - 30 = 2w + 3$

or $2w = 39 - 33 = 6$

or $w = 3$ and $w + 3 = 6.$

The strip of walk on the side opposite the sundial is six feet wide.

27. coconuts

This problem can be solved in a straightforward manner by letting x equal the amount each man receives when the final division is made and working back until an equation is found for y, the number of coconuts originally on the island. In terms of x

$$\text{the equation is } y = \frac{1024x + 1332}{81}.$$

This equation can be solved (by the method used in earlier problems in this section) without difficulty and the answer

$$y = 1024k + 247 \quad \text{and} \quad x = 81k + 18$$

obtained. The following line of reasoning will, however, give the same answer a bit more directly:

The fourth man left $12R$ coconuts since he left a pile which could be divided equally (with no remainder) into either three or four piles. He must have found, therefore, $16R + 3$ coconuts. It follows that $16R + 3$ must be divisible by 3 (the third man left three equal piles). Let R equal $3S$ and $16R + 3$ will become $48S + 3$; the third man must have taken $16S + 1$ as his share. There must have been $64S + 4 + 3 = 64S + 7$ when he arrived.

Hence, since this is the number of coconuts the second man left, it also must be divisible by 3. Let S equal $3T + 2$. Then the number the second man left is $192T + 135$. The amount he took would be $64T + 45$. Hence the amount the second man found (and the first man left) is $256T + 180 + 3 = 256T + 183$. This number must, naturally, be divisible by 3. If T is set equal to $3K$ this will be the case and it will equal $768K + 183$. Hence the first man would have taken $256K + 61$ and he would have found $1024K + 244 + 3 = 1024K + 247$.

It follows that $y = 1024K + 247$. To get the value of x is a simple matter of substitution:

$$x = 12R/4 = 3R = 3(3S) = 9S = 9(3T + 2)$$
$$= 27T + 18 = 27(3K) + 18$$
or $$x = 81K + 18.$$

By setting $k = 0$ the least number of coconuts that there could have been initially is 247.

In case you are interested in a general solution to the problem, here it is.

Assume an original pile of C coconuts and a group of n men. Each man in turn divides the coconuts he finds into n equal piles with a smaller number, say m, left over. He throws the leftovers to the monkeys, takes one pile for himself, and mixes the rest up. In the morning there are just enough coconuts left so that each man can take X.

Then
$$C = r \cdot n^{(n+1)} - m(-n)^n - m(n - 1)$$

where r is any integer so chosen that C will be positive. The smaller the value of r, the smaller the value of C. In a case where there are an odd number of men r can be any integral value from zero on up; in the case of an even number of men r cannot be zero but may be any integral value from one on up. In either case

$$X = \frac{[C + m(n - 1)][(n - 1)/n]^n - m(n - 1)}{n}.$$

28. a box of chocolates

Clearly Grandpa pays out, for each chocolate eaten, as many cents as there are children less one cent (not paid to the child who eats the chocolate), less three cents (deducted from the sum the eater of the chocolate would otherwise receive).

That is, if there are n children and c chocolates eaten,

$$(n - 4)c = 469 \quad (\$4.69 \text{ expressed as cents}).$$

The only factors of 469 are

$$1, 7, 67, 469,$$

so $(n - 4) = 1$ or 7 (67 and 469 are eliminated since each child ate at least one chocolate) and n (the number of children) is either 5 or 11.

But 5 is impossible; six children are mentioned by name.

Therefore, eleven children eat sixty-seven chocolates, and 67 can be partitioned into eleven different numbers in only one way:

$$1, 2, 3, 4, 5, 6, 7, 8, 9, 10, 12.$$

Eleven children in all ate chocolates; Edna ate twelve.

29. "seventeeners"

Let there be n tests and p girls participating, then the number of cents paid out by Miss Scribble per test is

$$17 - (p - 1) \quad \text{or} \quad (18 - p) \text{ cents.}$$

Since she paid out a total of 360 cents, it follows that

$$n(18 - p) = 360$$

where n and p are positive integers. The following table lists the possible solutions:

1. $p = 3$ $n = 24$
2. $p = 6$ $n = 30$
3. $p = 8$ $n = 36$
4. $p = 9$ $n = 40$
5. $p = 10$ $n = 45$
6. $p = 12$ $n = 60$
7. $p = 13$ $n = 72$
8. $p = 14$ $n = 90$
9. $p = 15$ $n = 120$
10. $p = 16$ $n = 180$
11. $p = 17$ $n = 360$

But of these possibilities:

1 to 5 inclusive are inadmissible. In these cases the maximum possible loss (which will be experienced by the girl who won only one test) would be $(n - 1) - 17$ and is less than the thirty cents Jane is known to have lost.

The number of tests must be equal to, or greater than,

$$1 + 2 + 3 + + + + p = \frac{p}{2}(p + 1)$$

if each girl is to win at least once and no two girls win the same number of tests. This condition rules out possibilities 6, 7, and 8.

Since the girls were in camp only two weeks and the number of tests did not exceed twelve per day on the average, the maximum number of tests is 12 times 14, or 168. This condition eliminates possibilities 10 and 11, leaving only possibility 9.

Hence there were fifteen participants and 120 tests, the number of tests won by the various girls being 1, 2, 3, . . . , 15.

If the number of tests Jane won is x, we have (since she lost a total of thirty cents)

$$(120 - x) \cdot 1 - x \cdot 17 = 30$$

or $$120 - 18x = 30$$

or $$x = 5.$$

Thus Jane won five of the 120 tests given.

A CATALOG OF SELECTED
DOVER BOOKS
IN ALL FIELDS OF INTEREST

A CATALOG OF SELECTED DOVER
BOOKS IN ALL FIELDS OF INTEREST

CONCERNING THE SPIRITUAL IN ART, Wassily Kandinsky. Pioneering work by father of abstract art. Thoughts on color theory, nature of art. Analysis of earlier masters. 12 illustrations. 80pp. of text. 5⅜ x 8½. 23411-8

ANIMALS: 1,419 Copyright-Free Illustrations of Mammals, Birds, Fish, Insects, etc., Jim Harter (ed.). Clear wood engravings present, in extremely lifelike poses, over 1,000 species of animals. One of the most extensive pictorial sourcebooks of its kind. Captions. Index. 284pp. 9 x 12. 23766-4

CELTIC ART: The Methods of Construction, George Bain. Simple geometric techniques for making Celtic interlacements, spirals, Kells-type initials, animals, humans, etc. Over 500 illustrations. 160pp. 9 x 12. (Available in U.S. only.) 22923-8

AN ATLAS OF ANATOMY FOR ARTISTS, Fritz Schider. Most thorough reference work on art anatomy in the world. Hundreds of illustrations, including selections from works by Vesalius, Leonardo, Goya, Ingres, Michelangelo, others. 593 illustrations. 192pp. 7⅛ x 10¼. 20241-0

CELTIC HAND STROKE-BY-STROKE (Irish Half-Uncial from "The Book of Kells"): An Arthur Baker Calligraphy Manual, Arthur Baker. Complete guide to creating each letter of the alphabet in distinctive Celtic manner. Covers hand position, strokes, pens, inks, paper, more. Illustrated. 48pp. 8¼ x 11. 24336-2

EASY ORIGAMI, John Montroll. Charming collection of 32 projects (hat, cup, pelican, piano, swan, many more) specially designed for the novice origami hobbyist. Clearly illustrated easy-to-follow instructions insure that even beginning papercrafters will achieve successful results. 48pp. 8¼ x 11. 27298-2

THE COMPLETE BOOK OF BIRDHOUSE CONSTRUCTION FOR WOOD-WORKERS, Scott D. Campbell. Detailed instructions, illustrations, tables. Also data on bird habitat and instinct patterns. Bibliography. 3 tables. 63 illustrations in 15 figures. 48pp. 5¼ x 8½. 24407-5

BLOOMINGDALE'S ILLUSTRATED 1886 CATALOG: Fashions, Dry Goods and Housewares, Bloomingdale Brothers. Famed merchants' extremely rare catalog depicting about 1,700 products: clothing, housewares, firearms, dry goods, jewelry, more. Invaluable for dating, identifying vintage items. Also, copyright-free graphics for artists, designers. Co-published with Henry Ford Museum & Greenfield Village. 160pp. 8¼ x 11. 25780-0

HISTORIC COSTUME IN PICTURES, Braun & Schneider. Over 1,450 costumed figures in clearly detailed engravings–from dawn of civilization to end of 19th century. Captions. Many folk costumes. 256pp. 8⅜ x 11¼. 23150-X

CATALOG OF DOVER BOOKS

THE STORY OF THE TITANIC AS TOLD BY ITS SURVIVORS, Jack Winocour (ed.). What it was really like. Panic, despair, shocking inefficiency, and a little heroism. More thrilling than any fictional account. 26 illustrations. 320pp. 5⅜ x 8½. 20610-6

FAIRY AND FOLK TALES OF THE IRISH PEASANTRY, William Butler Yeats (ed.). Treasury of 64 tales from the twilight world of Celtic myth and legend: "The Soul Cages," "The Kildare Pooka," "King O'Toole and his Goose," many more. Introduction and Notes by W. B. Yeats. 352pp. 5⅜ x 8½. 26941-8

BUDDHIST MAHAYANA TEXTS, E. B. Cowell and others (eds.). Superb, accurate translations of basic documents in Mahayana Buddhism, highly important in history of religions. The Buddha-karita of Asvaghosha, Larger Sukhavativyuha, more. 448pp. 5⅜ x 8½. 25552-2

ONE TWO THREE . . . INFINITY: Facts and Speculations of Science, George Gamow. Great physicist's fascinating, readable overview of contemporary science: number theory, relativity, fourth dimension, entropy, genes, atomic structure, much more. 128 illustrations. Index. 352pp. 5⅜ x 8½. 25664-2

EXPERIMENTATION AND MEASUREMENT, W. J. Youden. Introductory manual explains laws of measurement in simple terms and offers tips for achieving accuracy and minimizing errors. Mathematics of measurement, use of instruments, experimenting with machines. 1994 edition. Foreword. Preface. Introduction. Epilogue. Selected Readings. Glossary. Index. Tables and figures. 128pp. 5⅜ x 8½. 40451-X

DALÍ ON MODERN ART: The Cuckolds of Antiquated Modern Art, Salvador Dalí. Influential painter skewers modern art and its practitioners. Outrageous evaluations of Picasso, Cézanne, Turner, more. 15 renderings of paintings discussed. 44 calligraphic decorations by Dalí. 96pp. 5⅜ x 8½. (Available in U.S. only.) 29220-7

ANTIQUE PLAYING CARDS: A Pictorial History, Henry René D'Allemagne. Over 900 elaborate, decorative images from rare playing cards (14th–20th centuries): Bacchus, death, dancing dogs, hunting scenes, royal coats of arms, players cheating, much more. 96pp. 9¼ x 12¼. 29265-7

MAKING FURNITURE MASTERPIECES: 30 Projects with Measured Drawings, Franklin H. Gottshall. Step-by-step instructions, illustrations for constructing handsome, useful pieces, among them a Sheraton desk, Chippendale chair, Spanish desk, Queen Anne table and a William and Mary dressing mirror. 224pp. 8⅛ x 11¼. 29338-6

THE FOSSIL BOOK: A Record of Prehistoric Life, Patricia V. Rich et al. Profusely illustrated definitive guide covers everything from single-celled organisms and dinosaurs to birds and mammals and the interplay between climate and man. Over 1,500 illustrations. 760pp. 7½ x 10⅛. 29371-8

Paperbound unless otherwise indicated. Available at your book dealer, online at **www.doverpublications.com**, or by writing to Dept. GI, Dover Publications, Inc., 31 East 2nd Street, Mineola, NY 11501. For current price information or for free catalogues (please indicate field of interest), write to Dover Publications or log on to **www.doverpublications.com** and see every Dover book in print. Dover publishes more than 500 books each year on science, elementary and advanced mathematics, biology, music, art, literary history, social sciences, and other areas.